VISUALIZATIONS OF URBAN SPACE

This book explores environments where art, imagination, and creative practice meet urban spaces at the point where they connect to the digital world. It investigates relationships between urban visualizations, aesthetics, and politics in the context of new technologies, and social and urban challenges toward the Sustainable Development Goals.

Responding to questions stemming from critical theory, the book focuses on an interdisciplinary actualization of technological developments and social challenges. It demonstrates how art, architecture, and design can transform culture, society, and nature through artistic and cultural achievements, integration, and new developments. The book begins with the theoretical framework of social aesthetics theories before discussing global contemporary visual culture and technological evolution. Across the 12 chapters, it looks at how architecture and design play significant roles in causing and solving complex environmental transformations in the digital turn. By fostering transdisciplinary encounters between architecture, design, visual arts, and cinematography, this book presents different theoretical approaches to how the arts' interplay with the environment responds to the logic of the constructions of reality.

This book will appeal to scholars, researchers, and upper-level students in aesthetics, philosophy, visual cultural studies, communication studies, and media studies with a particular interest in sociopolitical and environmental discussions.

Christiane Wagner is a visiting research professor at the Contemporary Art Museum of the University of São Paulo (MAC USP), Brazil. Her research focuses on culture, media, education, image studies, and the interface of architecture, art, design, and digital technology concerning cultural heritage, politics, postcolonial aesthetics, and urbanization. She is a member of the Scientific Committee of the Society for Interdisciplinary Image Science in Germany. Wagner is also a member of the International Committee and chair of the International Publications Subcommittee of the College Art Association of America (CAA) New York.

Advances in Urban Sustainability

Urban Sustainability in Theory and Practice

Urban Sustainability in Theory and Practice
Circles of Sustainability
Paul James

Sustainability Citizenship in Cities
Theory and Practice
Edited by Ralph Horne, John Fien, Beau B. Beza, and Anitra Nelson

Street Fights in Copenhagen
Bicycle and Car Politics in a Green Mobility City
Jason Henderson and Natalie Marie Gulsrud

Adaptation Urbanism and Resilient Communities
Transforming Streets to Address Climate Change
Billy Fields and John L. Renne

Performance Metrics for Sustainable Cities
Edited by Sylvie Albert and Manish Pandey

The City is an Ecosystem
Sustainable Education, Policy, and Practice
Edited by Deborah Mutnick, Margaret Cuonzo, Carole Griffiths, Timothy Leslie, and Jay M. Shuttleworth

Visualizations of Urban Space
Digital Age, Aesthetics, and Politics
Christiane Wagner

For more information about this series, please visit: www.routledge.com/Advances-in-Urban-Sustainability/book-series/AUS

VISUALIZATIONS OF URBAN SPACE

Digital Age, Aesthetics, and Politics

Christiane Wagner

Routledge
Taylor & Francis Group
LONDON AND NEW YORK

earthscan
from Routledge

Designed cover image: Christiane Wagner

First published 2023
by Routledge
4 Park Square, Milton Park, Abingdon, Oxon OX14 4RN

and by Routledge
605 Third Avenue, New York, NY 10158

Routledge is an imprint of the Taylor & Francis Group, an informa business

British Library Cataloguing-in-Publication Data
A catalogue record for this book is available from the British Library

ISBN: 978-1-032-32417-3 (hbk)
ISBN: 978-1-032-32418-0 (pbk)
ISBN: 978-1-003-31494-3 (ebk)

DOI: 10.4324/9781003314943

Typeset in Bembo
by codeMantra

For Nikolas, my son

CONTENTS

FOREWORD

By teaching us to actively perceive what we passively see, to understand at deeper levels the life whose surface alone our eyes record, Christiane Wagner's *Visualizations of Urban Space: Digital Age, Aesthetics, and Politics* undertakes the timely, invaluable work of illumination. *Visualizations* reveals the meaning and significance of images that are the irrepressible offspring of the continuous interactions between metropolitan habitats, their omniform aesthetics, cultural processes, and broadband social media. Wagner's contribution is to discover within this open-ended network of evolving interconnections powerful new sources for enlightenment that are able to make us aware of immanent and potentially imminent possibilities for world-historical democratic social change. Such change changes everything. For our world, at our historical point in time, changing everything means nothing less than a planetary future newly populated by an all-inclusive community of social differences who respect and protect the natural world on which it depends for its every breath. To achieve those ends, *Visualizations* shows that image and sustainability must be taught together.

As I have no doubt Wagner's remarkable book will inspire a like-minded super-abundance of uplifting thoughts and feelings, I want to share my own first reaction to her work. I realized I had been waiting for "this book," one whose author had the breadth and depth, theoretical imagination, and clear understanding of what can be done to nurture a public sphere devoted to the political possibility of a sustainable environment. It is an incipient public sphere that we—earth's inhabitants—are already always engaged in that offers the promise of a critical consciousness able to redirect our form of life to different ends. It is that consciousness that Wagner seeks to educate, while the different ends to which her pedagogy leads culminate in a sustainable future. To Wagner, nothing is more important. Nothing more possible.

Morton Schoolman
Professor of Political Science
State University of New York at Albany

INTRODUCTION

Visualizing urban space refers to the faculty of apprehension through the senses. That is, it means to perceive the city environment through visual stimuli. Specifically, visual perception is the ability to see and interpret clear information from the visual spectrum. The action of visualizing is a transformation of concepts into real or mentally visible images. One must contend with two kinds of images to think about visualizing: The first is objectively understood—a graphic, photographic, plastic, or dynamic visual representation; the second is subjective—an idea that one conceptually has of something. Images exceed physical and knowledge boundaries through imagination; and create and provide free access to creativity, and thus, have a place beyond the immediate social environment. Further to this preliminary definition, space perception is added, which is the ability to understand the relationship between human beings and the environment that they occupy.

Perception is an active process of selecting, organizing, and interpreting information. According to Rudolf Arnheim (1977), perceiving formal patterns enables similar aesthetic responses for individuals from different cultures. Therefore, between different cultures and worldviews, there can be a consensus on urban configurations. As the practicing architect Ralf Weber (1995) points out, with advances in neuroscience, many urban projects are applying empirical aesthetics, evaluating the perception process through scientific methods to identify an aesthetic consensus for urban planning with a focus on the environmental quality of citizens' daily lives. Empirical aesthetics deals with the visual perception of the architectural composition of buildings, encompassing formal aesthetics and the associations established with the physical and natural elements in an urban scene as a result of the cognition process under symbolic aesthetics.

Nevertheless, understanding the dynamics of urban space in its complexity as an essential part of everyday life is to realize that the fundamental bases lie in the

DOI: 10.4324/9781003314943-1

construction of space by society and culture, and also in the political and economic aspects that concern an aesthetic consensus. The social city's structure and the arts in all its cultural manifestations define the perception of its inhabitants. Therefore, equally important is philosophical aesthetics in the face of advances in cognitive science. Likewise, the Kantian aesthetic consensus in its communication process should be overcome. For Kant, the dependence on the conception of space and time would be in the mind as pure forms of intuition and nonempirical representation. Consequently, the possible common interests as well as the conception of space and time, independent or not from experience of the things of the world, are contemporarily discussed. Hence, ontologically, whether the notions of space and time depend on their relations to objects or not, from Leibniz and Newton to Kant's considerations and the origin of the representation of what defines Euclidean geometry, this book seeks to overcome Kant's *Transcendental Aesthetics* (1790) regarding a purely *a priori* conception of space and time.

Furthermore, beyond this complexity of understanding the visual and mental experience regarding space and time, the present work focuses on the construction of space related to time and the presence of things and humans in their historical, social, and cultural values as essential to the associations established in urban scenes, in addition to the Hegelian spirit of the time—namely *Zeitgeist*. In addition, the analysis considers the conception of social space defined by Henri Lefebvre (1974) when establishing a social vision—which is in opposition to the abstract and geometric principles of Euclidean space and Cartesian philosophy—to develop a relationship between the mathematical logic of space and the practical mental and social world. That is, an analysis is established between the philosophical and geometrical space for the space of everyday life. Through this classical theoretical contextualization, *Visualizations of Urban Space: Digital Age, Aesthetics, and Politics* seeks to update the context of the everyday experience of metropolises through an interdisciplinary analysis in philosophy, social sciences, humanities, and arts.

Consequently, for an aesthetic, cultural, social, and political approach to urban visualization, it is necessary to consider the visual manifestation in its various forms of realization in public space. From the earliest civilizations, through the classical Greek and Roman representations, the Middle Ages, the Renaissance, and modernity, to the present day, many references have visually configured the public space. Specifically, art, as much as it is understood as a representation technique in content and form, is simultaneously the fruit of individual and collective creativity to be presented and communicated through the public's participation. Before the current media, such as cinema, video, television, photography, and digital technologies, mediated visual configurations throughout the World Wide Web, the first forms of presentation in antiquity were found directly in the streets and city centers. Therefore, the most traditional form of aesthetic experience is not found in museums, galleries, libraries, or cultural institutions through photographs, films, or drawn and cataloged records. Even if these records are also ways of making art, crucial here is to understand that the characteristic of the visual experience in urban space is what is essential. Even before these institutionalized places were considered appropriate for the appreciation and exhibition of artistic heritage, art already existed in its direct connection with everyday life in the squares, streets, and the city's shared spaces.

Equally critical is the discussion of the effect of utopias concerning sociopolitical, historical, and technological movements and culture and their relationships with human nature and the construction of public space. The metropolis's structures are described by discussing public space and arriving at an interpretation of Morton Schoolman's work (2020), where the democratic enlightenment and aesthetics of the private sphere are compared to understand their effects on public spaces. Therefore, the focus is on an image moving, emphasizing the urban visual structures in motion—or visual narratives. These images portray the sociocultural context through people's capacity to create narratives that configure the collective consciousness and shape public opinion. Moreover, this book's analysis aims to bring the city visual experience closer to social reality as a cultural value in contemporary days and an (in)tangible heritage of humanity, also represented in the virtual space. Thus, regarding the article "Is the City Becoming Computable?" by Antoine Picon and Thomas Shay Hill, the present analysis supposes that today there is no doubt that urban public space has expanded to the Web, enabling new forms of visualization from the urban physical space to become digital as well as integrating the virtual space to allow interactions.

To extend the cultural, aesthetic, and political aspects of this approach to contemporary urban visualizations, the first chapter, titled "Cultural Images: Real and Imagined Experiences," delimits the image's meaning as real and imagined experiences to understand the power of narratives and media in consumer society. Therefore, the chapter analyzes the polysemic effect of media images and exposure value. The focus is on the art of representation, simulation, and dissimulation concerning the imagined reality of social actors in their limits (Benjamin 1933, 1939; Debord 1967; Baudrillard 1981). Ethically and aesthetically, the social reality experienced by appearances is analyzed in direct relation to the information society. The analysis explores the social patterns highlighted in the arts and digital media universe, emphasizing social networks. Accordingly, two fundamental aspects direct the discussion. The first aspect is the ephemeral image in constant reconfiguration by digital media and interactive networks, which is to the detriment of intensive and individual experiences. The second aspect is a focus on the transforming aesthetic, which consequently reconfigures already established cultural values into new forms of representations. In short, the discussion concerns the image in the face of immediate and superficial experiences of collective exposure toward possibilities of individual resilience (Nissenbaum 2009; Buchanan and Zimmer 2018). Following this analysis of the contemporary context with some references to modern times, the goal of the second chapter, "Urban Space and Collective Consciousness," is to deepen the discussion of the transformation process from the modern to the current context regarding public and urban spaces. The chapter aims to present a visual and theoretical analysis based on the ideals of a democratic society. The cultural paths of social, political, and economic history have always been transferred to art; therefore, the chapter analyzes the artistic realizations embedded in the ideological context of modern times to situate contemporary society (Adorno 1970; Habermas 1962, 1981), its global dynamics, and the effect of media concerning everyday aesthetics. Hence, in general, what constitutes the configuration of a contemporary urban space is discussed regarding the collective consciousness *vis-à-vis* the dynamics of global metropolises in their transformation process, leading to other reflections on the constant visualization process. It is a process of achievement within the limits of creation, regardless of the historical context and the idea of freedom bound to the entire transformation and development of cultures. Consequently, this chapter explores metropolises' structures and the social dynamics of functioning, organizing, and restructuring the possible image configurations produced by the cultural context as reality. Moreover, it concerns the collective consciousness and vision of the democratic ideals that guide society toward its fundamental rights, not only as an ideology or an image but also as a social reality (Bobbio 2007). These rights respond to the need to maintain the social order—a unity—with regard to the cultural diversity manifested in the arts and media. Hence, this chapter concerns the power of the images and visual narratives that represent the global culture's ideal of democracy (Schoolman 2020).

Following on from this theme, the purpose of the third chapter, titled "The Arts and Traces of the Ideal City," is to understand mimetic images as representations of reality. The fundamental axis of this reflection is the relationship between the artist's intention and the cultural production, which creates an "image" of reality in mutation. The objective and subjective values by which beauty and horror are displayed in public space—both as image and as art—must be viewed as visualizations that cross boundaries, creating and offering free access to the imagination through the media and in the urban space. Thus, images project beyond the immediate social environment. Nevertheless, visual transgression is challenged when the concrete image is used as an artistic, cultural, media, and expressive product concerning fundamental freedoms. The notion of freedom is seen as the ideal of beauty and perfection that surpasses everyday experiences, associated with divinity and the unachievable ideas of truth and justice (Jäger 1947; Gurley 1999). Therefore, the discussion highlights that thinking of art, culture, and reality in public space with an emphasis on aesthetic theories implies Western and non-Western knowledge (Auerbach 1994; Rampnoux 2008). In this direction, in the fourth chapter, titled "Democratization of Art and Culture," some essential observations about the arts as a means of communication are presented. Thus, writing and verbal languages have developed and are, in their technical evolutions, the main drivers of communications, societies, and cultures. Since then, our knowledge of the evolution of communications processes has been visually represented in development and interpretation (Meggs and Purvis 2016). Moreover, imagination drives creativity and the development of techniques to overcome the limits of our language, of the idea's representation, and of more understanding, knowledge, and (mainly) the human capacity to more effectively learn, communicate, and relate to the environment. Furthermore, the chapter discusses the interdependent system in which social origin and educational differences are characterized (Bourdieu 1998). The most significant aspect is the cultural legacy of the development of visual and written languages in terms of the evolution of arts and communication with regard to society, culture, and, today, cultural institutions. Generally, the dynamics of the art of publication and art as content aimed at understanding the appropriation of public and urban space are discussed to contextualize the need for techniques, processes, planning, and employment of the digital technologies expanding public and urban spaces (Grabbe 2018; Hallin 2020).

In the fifth chapter, titled "Visual Media Effects," the repercussion of photography and the role of the media, photojournalism, and the moment of capturing the image through visual appeal in the formal aspects are highlighted. An analogy with the oldest representations of freedom and victory in human history through artistic forms is immediate. In recent decades, the media has amplified these essential references in the mimetic representation of manifestations of fundamental human values. In this process of expression (*mimesis*), as far as feelings are concerned, the visual representations manage to free themselves from everyday relationships. This chapter deals with human beings' paradoxes, complexities, and particularities through art and its visual effects in the media and public space. Thus, in the face of the practical process of applying art to images, both figurative and abstract, the role of the media stands out in the discussion by proposing experiences that condition individuals in their daily urban reality. Therefore, an iconographic analysis is conducted to discover the leitmotiv of freedom and victory as visual representation, under a moral and ethical basis, among different cultures (Panofsky 1955; Bourdieu 1991). In this sense, a critical approach to mass communication, literature, and the arts is presented in its media functions, which culminate in general visual culture and are also related to childhood (Piaget 1932, 1937, 1975). In this regard, given that images of children depict inequality, vulnerability, and injustice so often in the media, the need to recognize injustice and understand the developmental needs of children is evident. Hence, the chapter investigates the visual representation concerning the exploitative use of children's images to the detriment of their fundamental human rights or, more specifically, the rights of the child. Accordingly, the analysis describes signs related to the power of the image based on traditional Western cultures.

Thereafter, the sixth chapter, titled "Urban Resilience: Toward Postcolonial Aesthetics," focuses on postcolonial aesthetics to discuss the theoretical challenges that seek to overcome the generalized notions of modern European thinking through an investigation of non-Western literature and art that seek to coexist and overcome the hegemonic Western forms of globalization (Said 1978; During 1985; Caldas 1987; Chakrabarty 2000; Mbembe 2001; Bahri 2003; Mignolo and Walsh 2018; Appadurai 2021). However, the construction of modern knowledge and the sense of Western progress are discussed as the major challenges in decolonization, considering that metropolises and colonies are co-constitutive elements of a growing world capitalist order of development and civilization (Comaroff 2012). For this, sociocultural, political, ethical, and aesthetic values are discussed, especially the cultural values of native peoples and afro-descendants in contemporary times toward the digital turn (Canevacci 2004, 2011, 2022). In this direction, some factors are essential to consider, such as aesthetic appreciation and the notion of artistic beauty and its reverse focused on the current context, considering cultural differences, especially the global South and indigenous and tribal peoples' cosmologies, Afropolitanism, and the Afro-diasporic movement.

Continuing from this postcolonial aesthetic notion, the seventh chapter, titled "Sociopolitical Facts, Art, and Technology," focuses on the aesthetic experience with the current urban visual scenario as a means of overcoming the construction of difference as otherness from Morton Schoolman's (2020) conception. To this end, facts related to the arts and technologies that configure the image of the urban environment are directed to the democratic enlightenment associated with the perception of private spaces with the public as visualizations in movement, rhythm, and a dynamic form of guiding reason. The discussion highlights the inventions and apparatus of cultural production and the technological advancement of the moving image through cinematographic production and the mental narrative of remissions and daily experiences to meet an aesthetic education. Cultural production is also discussed in its rational aspects based on the industrial impact of automation as a significant consequence of the perception of freedom established by contemporary society in visualizing the future (Weber 1922; Horkheimer and Adorno 1947; Wiener 1950; McLuhan 1964; Habermas 1968; Adorno 1970; Deleuze and Guattari 1972; Pinker 2018). Moving toward the future, the sense of technicization has transformed urban space aesthetically.

Consequently, through this aestheticization, the discussion in the eighth chapter, titled "Technicization and Aestheticization of Social and Urban Structures," is associated with the contemporary context, which involves the general and superficial aspects of aesthetic experience (Welsch 1996). The material world is discussed in its aspects of improvement by technicization given the development of science and technology in the face of capitalism and its forms of commodification, which configure a superficial environment that involves the entire social and public sphere. Moreover, the four phases of the industrial revolution are addressed, leading to the present moment of the fourth revolution and the implications in the urban context under the digital shift. This digital context comprises robotics, artificial intelligence (AI), machine-to-machine connections, big data, the Internet of Things, cyber security, 3D modeling, and cloud computing as urban life patterns. Nevertheless, considering the changes in urban space (Picon 2005–2006), the liberal market in its course of industrial development implies the technicization and aestheticization of social and urban structures. In this sense, the articulations focused on using digital technologies are discussed in terms of their advantages and disadvantages as well as politically

(Languillon-Aussel 2021). Moreover, the concept of the appropriation of public space (Lefebvre 1968) stands out through the exercise of citizenship, which refers to the inclusion of all urban inhabitants through the right to participation and appropriation. Through such participation, urban citizens can access the decisions that produce the urban space. Thus, social inclusion projects for achieving essential Sustainable Development Goals are analyzed as solutions to inequalities and housing problems, and even the situation of cities under the tutelage of digital technologies.

Based on the considerations that form the chapters thus far, the ninth chapter, titled "The Urban Image in Motion and Connection," analyzes the technical effects of the moving image and architectural and urban structures on the interaction of physical and virtual spaces by digital technologies considering democratic values. The camera represents the main element in this connection and interaction through capturing images of the human presence in its activities. Consequently, the analysis focuses on the technical and ethical aspects of image capture. Furthermore, the necessary conditions are discussed for the state's rights as public power as well as the rights of the private sector in the practices that control public access. These practices range from the capture of images to the applications of AI in automation processes surrounding the functioning of an urban city, in their behavioral implications regarding everyday experiences, and in the sense of a smart city. Consequently, using digital technologies creates new and challenging opportunities in the field of sustainability (Friedman 2021), and visual media enables the enhanced communication of structures for improved results in urban design (Batty 2013). Therefore, the context of the digitalization of cities is discussed under the aspects of internal and external perception as well as how complex digital systems make such digitalization possible to simulate social phenomena involving time and space as heritage, that is, the fact that historical circumstances matter (Picon 2015; Picon and Hill 2020).

Equally crucial for a closer approximation of the role of the arts, the tenth chapter, titled "The Arts' Function: Social Inclusion and Sustainability," considers that the city's history is essentially the history of art (Argan 1983). This historical aspect is the basis for constructing and reconstructing the predominant factors that shape the urban environment in its transformation process of visualizing the future. In this sense, the emphasis is on the future of cities concerning the past and how investments in cultural heritage preservation and urban planning are conducted (Holtorf and Högberg 2021). This implies, as far as historical facts are concerned, the past and the construction of the future. Then, the emerging concern in the arts related to environmental awareness and the need for sustainable development in the face of climate change are addressed, emphasizing the interdependence between architecture, art, and design in their technical essence to seek social solutions. To this end, contemporary challenges aimed at social equality and greenhouse gas emission reduction are discussed with an emphasis on design (Whiteley 1993; Bürdek 2015), as are the risks to society concerning climate change and the search for awareness through the arts in their critical and revolutionary forms (Welsch 2019; Beck [1986] 1993). The aim is to present an aesthetic-social analysis and understand the reciprocal relationship between the arts, the current environment, urban challenges, and political will.

Subsequently, in the eleventh chapter, titled "Designing for Sustainability," the focus is on historical, theoretical, and design education related to the goals of sustainability and the challenges of design for solving problems (Bürdek 2015, 2019; Papanek [1971] 2019). Nonetheless, the context of corporate image is considered, in which design plays a key role. Moreover, the challenges for planning and developing products and services related to social goals and conceptions for building a corporate image based on problem-solving are discussed. The solutions to these problems involve everything from risk reduction to improved strategies for meeting the Sustainable Development Goals. Thus, the creative process is discussed in its potential for innovation and social transformation in which design contributes to economic, scientific, and technological development, which brings us to the end of the book. In the twelfth and final chapter, titled "Aesthetic, Social, and Urban Solutions for a New World View," everyday aesthetics is related to the growing presence of the new technologies and the digital world in the urban lifestyle and its cultural and political effects. In this regard, the meaning of "aesthetics," associated with daily experiences, ethically and aesthetically concerns individuals' perceptions of things in their urban environment (Berleant 1984; Carlson 2008, 2010, 2020). To this end, a discussion is provided of climate change and environmental issues in the social context of awareness. The technological revolution had an early impact on urban life and consequently on the values of "romantic nature" as opposed to industrial development, and it is addressed in discussions of the worldviews and utopias based on a "back to nature" vision (Morton 2007). Next, sustainability is analyzed in its economic dimension, focusing on the social context of the emerging policy agenda as the main challenge for corporations in the 21st century, which respectively concerns design challenges (Friedman 2021). Finally, the Sustainable Development Goals are discussed in terms of their complexity concerning the responsibilities of long-term planning (Elkington 1997; The New Urban Agenda 2016; Friedman 2021). This planning is discussed under the conditions of public policy's ethical and aesthetic aspects and the population's needs. Finally, this book addresses the formulation of an enhanced understanding of the complexity of the urban environment—that is, of urban lives—in social-spatial configurations characterized by the interplay among physical and digital realities toward sustainability.

Bibliography

Adorno, Theodor. *Ästhetische Theorie* [Aesthetic Theory]. Frankfurt am Main: Suhrkamp, 1970.

Althusser, Louis. "Idéologie et appareils idéologiques d'État (Notes pour une recherche)" [Ideology and Ideological State Apparatus (Notes for a search)]. *La Pensée* no. 151 (1970). Reprint, in *Positions*, 1964–1975. Paris: Les Éditions Sociales, 1976.

Appadurai, Arjun. "Beyond Domination: The Future and Past of Decolonization." *The Nation*, 9 March 2021. https://www.thenation.com/article/world/achille-mbembe-walter-mignolo-catherine-walsh-decolonization/

Argan, Giulio Carlo. *Storia dell'arte come storia della città* [Art History as the History of the City]. Rome: Editori Riuniti, 1983.

Arnheim, Rudolf. *The Dynamics of Architectural Form*. Berkeley: University of California Press, 1977.

Auerbach, Erich. *Mimesis: dargestellte Wirklichkeit in der abendländischen Literatur*. 1946. Reprint, Basel: Francke, 1994. *Mimesis: The Representation of Reality in Western Literature*. Trans. by Willard R. Trask. Princeton, NJ and Oxford: Princeton University Press, 2003.

Batty, Michael. *The New Science of Cities*. Cambridge, MA: MIT Press, 2013.

Baudrillard, Jean. *Simulacra and Simulation*. Trans. by Sheila F. Glaser. Michigan: University of Michigan Press, 1994. Originally published as *Simulacres et simulation* (Paris: Galilée, 1981).

Baudrillard, Jean. *La société de consommation: ses mythes, ses structures* [The Consumer Society: Myths and Structures]. Paris: Gallimard, 1997. Reprint, London: Sage, 1999.

Beck, Ulrich. *Risk Society: Towards a New Modernity*. 1986. London: Sage, 1993.

Benjamin, Walter. "Experience and Poverty." In *Die Welt im Wort* (Prague). Gesammelte Schriften, II. Trans. by Rodney Livingstone (December 1933): 2013–2019. Reprint, *Walter Benjamin: Selected Writings 1927–1934*, edited by Michael W. Jennings, Howard Eiland, and Gary Smith, trans. by Rodney Livingstone and others, 731–736. Cambridge, MA: The Belknap Press of Harvard University Press, 1999.

Benjamin, Walter. "Das Kunstwerk im Zeitalter seiner technischen Reproduzierbarkeit" [The Work of Art in the Age of Mechanical Reproduction]. 1939. Reprint, *Medienästhetische Schriften*. Frankfurt am Main: Suhrkamp, 2002.

Berleant, Arnold. "Aesthetic Participation and the Urban Environment." *Urban Resources* 1 (1984): 37–42.

Bobbio, Norberto. *Le futur de la démocratie* [The Future of Democracy]. Paris: Seuil, 2007.

Bourdieu, Pierre. *Language and Symbolic Power*. Cambridge: Polity Press, 1991.

Bourdieu, Pierre. *Les règles de l'art* [The Rules of Art]. Paris: Éditions du Seuil, 1998.

Buchanan, Elizabeth A., and Michael Zimmer. "Internet Research Ethics." *The Stanford Encyclopedia of Philosophy* (Winter 2018). Edward N. Zalta, ed. https://plato.stanford.edu/archives/win2018/entries/ethics-internet-research/

Bürdek, Bernhard E. *Design: History, Theory and Practice of Product Design*. Basel: Birkhäuser, 2015.

Caldas, Waldenyr. *A literatura da cultura de massa: uma análise sociológica* [The Literature of Mass Culture: A Sociological Analysis]. São Paulo: Lua Nova, 1987.

Canevacci, Massimo. *A cidade polifônica: ensaio sobre a antropologia da comunicação urbana* [The Polyphonic City: Essay on the Anthropology of Urban Communication]. 2004. Reprint, São Paulo: Studio Nobel, 2011.

Canevacci, Massimo. "Ubiquities: Aesthetic and Anthropology Glances on Undisciplined Identities." *Art Style, Art & Culture International Magazine* v. 9, no. 9, (March 2022): 11–25. doi: 10.5281/zenodo.6338582

Carlson, Allen. *Nature and Landscape: An Introduction to Environmental Aesthetics*. New York: Columbia University Press, 2008.

Carlson, Allen. "Contemporary Environmental Aesthetics and the Requirements of Environmentalism." *Environmental Values* 19 (2010): 289–314.

Carlson, Allen. "Environmental Aesthetics." *The Stanford Encyclopedia of Philosophy* (Winter 2020), Edward N. Zalta. https://plato.stanford.edu/archives/win2020/entries/environmental-aesthetics/

Chakrabarty, Dipesh. *Provincializing Europe: Postcolonial Thought and Historical Difference*. Princeton, NJ: Princeton UP, 2000.
Comaroff, Jean, and John L. Comaroff. *Introduction. Theory from the South: Or, How Euro-America Is Evolving Toward Africa*. New York: Paradigm Publishers, 2012.
Debord, Guy. *La société du spectacle* [Society of the Apectacle]. Paris: Gallimard, 1967.
Elkington, John. *Cannibals with Forks: The Triple Bottom Line of 21st Century Business*. Oxford: Capstone Publishing Limited, Oxford Centre for Innovation, 1997.
Francastel, Pierre. *Art et technique aux XIXe et XXe siècles* [Art and Technology in the 19th and 20th Centuries]. Paris: Gallimard, 1956.
Friedman, Avi. *Fundamentals of Sustainable Urban Design*. Cham: Springer, 2021.
Grabbe, Lars Christian. "Image Transformation: The Hyperaisthesis of Digital Images." In *Yearbook of Moving Image Studies, Image Evolution: Technological Transformations of Visual Media Culture*, edited by Lars Christian Grabbe, Patrick Rupert-Kruse, and Norbert M. Schmitz, 78–99. Marburg: Büchner-Verlag, 2018.
Gurley, Jennifer. "Platonic Paideia." *Philosophy and Literature* 23, no. 2 (1999): 351–377. Project Muse. https://doi.org/10.1353/phl.1999.0040
Habermas, Jürgen. *The Structural Transformation of the Public Sphere: An Inquiry into a Category of Bourgeois Society*. Trans. by Thomas Burger with the assitance of Frederick Lawrence. Cambridge, MA: MIT Press, 1989. Originally published as *Strukturwandel der Öffentlichkeit. Untersuchungen zu einer Kategorie der bürgerlichen Gesellschaft* (München: Luchterhand, 1962).
Habermas, Jürgen. *Technik und Wissenschaft als Ideologie* [Technology and Science as Ideology]. Frankfurt am Main: Suhrkamp, 1968.
Habermas, Jürgen. *Theory of Communicative Action*. Trans. by Thomas McCarthy. Boston: Beacon Press, 1984. Originally published as *Theorie des kommunikativen Handelns* (Frankfurt am Main: Suhrkamp, 1981).
Hegel, Georg Wilhelm Friedrich. *Lectures on the Philosophy of Art*. The Hotho Transcript of the 1823 Berlin Lectures. Trans. by R.F. Brown. Oxford: Clarendon Press, 2014. Originally published as *Vorlesung über die Philosophie der Kunst*. Berlin 1823. Rewritten by Heinrich Gustav Hotho (Hamburg: Felix Meiner, 1998).
Holtorf, Cornelius, and Anders Högberg, eds. *Cultural Heritage and the Future*. London and New York: Routledge, 2021.
Horkheimer, Max, and Theodor Adorno. *Dialektik der Aufklärung* [Dialectic of Enlightenment]. Amsterdam: Querido, 1947. Reprint, Frankfurt am Main: Fischer, 1998.
Kant, Immanuel. *Kritik der reinen Vernunft* [Critique of pure reason]. Riga: Johann Friedrich Hartknoch, 1781, 1787, 1790. Reprint, Stuttgart: Reclam Verlag, 2006.
Lefebvre, Henri. *Right to the City*. 1968. Trans. by Kofman, E. and Lebas, E., eds., *Writings on Cities*. Oxford: Blackwell, 1996.
Lefebvre, Henri. *The Production of Space*. Trans. by Donald Nicholson-Smith. Oxford: Blackwell, 1991. Originally published as *La production de l'espace* (Paris: Éditions Anthropos, 1974).
Mbembe, Achille. *On the Postcolony*. Berkeley: University of California Press, 2001.
McLuhan, Marshall. *Understanding Media: The Extensions of Man*. New York: McGraw-Hill, 1964.
Meggs, B. Philip, and Alston Purvis. *A History of Graphic Design*. Michigan: John Wiley & Sons, 2016.
Mignolo, Walter, and Catherine Walsh. *On Decoloniality: Concepts, Analytics, Praxis*. Durhan e Londres: Duke University Press, 2018.
Morton, Timothy. *Ecology without Nature: Rethinking Environmental Aesthetics*. Cambridge, MA and London: Harvard University Press, 2007.

Nissenbaum, Helen. *Privacy in Context: Technology, Policy, and the Integrity of Social Life.* Stanford, CA: Stanford University Press, 2009.

Papanek, Victor. *Design for the Real World: Human Ecology and Social Change.* 1971. Reprint, New York: Thames & Hudson, 2019.

Piaget, Jean. *Le jugement moral chez l'enfant* [Moral Judgment in Children]. Bibliothèque philosophie contemporaine. Paris: PUF, 1932.

Piaget, Jean. *La construction du réel chez l'enfant* [The Construction of Reality in Children]. Paris: Delachaux et Niestlé, 1937.

Piaget, Jean. *L'équilibration des structures cognitives: Problème central du développement* [Balancing Cognitive Structures: A Central Problem in Development]. Paris: PUF, 1975.

Picon, Antoine. "Construction History: Between Technological and Cultural History." *Construction History* 21, (2005–2006): 5–19. http://www.constructionhistory.co.uk/shop/

Picon, Antoine. *Smart Cities: A Spatialised Intelligence.* Chichester, West Sussex: Wiley, 2015.

Picon, Antoine, and Thomas Shay Hill. "Is the City Becoming Computable?" In *Architecture and the Smart City,* edited by Sergio M. Figueiredo, Sukanya Krishnamurthy, and Torsten Schroeder, 29–42. London and New York: Routledge, 2020.

Rampnoux, René. *Histoire de la pensée occidentale: de Socrate à Sartre* [History of Western Thought: From Socrates to Sartre]. Paris: Ellipses, 2008.

Schoolman, Morton. *A Democratic Enlightenment: The Reconciliation Image, Aesthetic Education, Possible Politics.* New York: Duke University Press, 2020.

Weber, Max. *Wirtschaft und Gesellschaft: Grundriss der verstehenden Soziologie* [Economy and Society: Outline of Understanding Sociology]. Berlin: Peter Kietzmann, 1922. https://www.textlog.de/weber_wirtschaft.html

Weber, Ralf. *On the Aesthetics of Architecture: A Psychological Approach to the Structure and the Order of Perceived Architectural Space.* Aldershot: Hampshire Avebury, 1995.

Welsch, Wolfgang. "Aestheticization Processes: Phenomena, Distinctions, and Prospects." *Theory, Culture & Society* 13, no. 1 (February 1996): 1–24. https://doi.org/10.1177/026327696013001001

Welsch, Wolfgang. "Art Addressing the Anthropocene." *Serbian Architectural Journal* 11, no. 2 (2019): 341–58. https://doi.org/10.5937/SAJ1902341W

Wiener, Norbert. *The Human Use of Human Beings: Cybernetics and Society.* Boston, MA: Houghton Mifflin, 1950.

1

CULTURAL IMAGES

Real and Imagined Experiences

The new experience permeates all aspects of everyday life, transforming it even before it can present its concrete realizations for the self-consciousness of the sociocultural condition. Cultural images offer sensitive aspects—representations—of humans' experience in their environment. They concern aesthetic values in art's formal elements and, as content (Hegel [1823] 2014), the human realization for environmental and everyday aesthetics. The meaning would not be in the moment of execution but in the formal result of the image in its visual effect and narrative on the social environment. The fundamental axis lies in the relationship between technology and culture, construction and hegemonic maintenance, consequently identifying significant references for the ensuing complexities of media practices. These intensified experiences derive mainly from recent and constant digital technologies, representing historical, political, and social moments in their spatial locations as places from which one could act. The central meaning is to discuss the individual's relationship with their sociocultural environment through everyday aesthetics, thereby transcending visual experience through the digital world. However, these are new habits that have resulted from other necessary living conditions in society, such as political will and investments to make the Internet accessible to all. Such transformations are inevitable. Thus, engagement with history or politics does not form a discourse independent from the aesthetic discourse, nor even from the discourse of science and technology, if political activity is considered part of the aesthetic reality, since the sensitive aspects encompass the content and, consequently, the political issue. Thus, individuals are transformed in an environment that is not increased but rather modified by constant technological innovations, transcending time and space in terms of reality in virtual environments. Nevertheless, these individuals simultaneously require physical spaces for a better relationship with their social or private environment, which evidences a need for human nature. This need

DOI: 10.4324/9781003314943-2

offers us no doubts in this reality of the information society, in which information plays an increasingly central role in social, cultural, and political life and has the objectives of promoting technological innovation and making communication processes more agile and efficient by involving industry, research, science, and knowledge.

Images not only narrate our everyday perceptions of the environment but also lead us to the illusion of many realities (Benjamin 1933, 1939). From this complexity of human perceptual faculties, the discussion in this chapter aims to reflect on images—virtual and real—as technical aspects and effects of social imagination. As such, how is it possible to perceive this experience? Is it potentially unreal, or is there a real power of effects? Along these lines, the argumentation attempts to answer these questions based on the traditional conception of reality, centered on assumptions of world perceptions, of things themselves, of concepts, and of representations of that reality in space-time. In the following analyses, invention and artifice are considered humanity's capacity to produce illusory meanings and experiences. There are many productions for the public transmission of an imaginary universe, thus representing the simulation of realities (Debord 1967; Baudrillard 1981). Therefore, all values become necessary for leading individuals and establishing them as subjects of the private economic system concerning the public context. Consequently, the information and its new means for a sociocultural representation are essential. However, aware of a social objective, through technological advances, information and communication technologies are focused on as a choice in transition periods that would define new meaning in society (Nissenbaum 2009; Buchanan and Zimmer 2018). Accordingly, based on the principle that this is a new phenomenon related to sociocultural transformations, the inductive method of qualitative analysis—based on selected literature (Weber 1922; Horkheimer and Adorno 1947; Adorno 1970; Habermas 1962, 1968, 1981), specific concepts, and discussions already conceptualized from observed patterns—is used for a discussion of what the new possibilities represent in the face of technological development. The discussion is divided into two interdependent stages, namely analysis and interpretation.

First, a critical interpretation of modern and contemporary societies is conducted, which concerns ephemeral experiences at the expense of essential values of urban life regarding social, artistic, and media representations, especially social media. Second, the relationship between the individual and technology—urban and digital culture—is analyzed regarding the technical images in relation to the field of visual culture. Finally, in this process of the interdependence of approaches between communication, art, and technology, technological innovation is considered. This innovation is characterized by the application of knowledge, information processing, and communication devices, in the function of an immediate utility in the organization, in which digital technology plays the leading role. An example is social media as an integral part of new sociocultural behaviors, which have, in turn, been discussed in recent studies on ethics and the integrity of contemporary social and urban life.

The Reality of an Enchanted World

Images convey fantasy values, creating a stimulus for consumption and the visual culture of contemporaneity. The influence of cultural production by innovation is the way images are configured. Cultural images are the visual environment in the face of polysemy. Creativity and technique are the means to accomplish novelty by believing in the image, designating the path through artifices for communication; moreover, experience has demonstrated how to find the perfect form: By observing repetition until one finds the expressive forms through which it is convenient to express the ideas to spectators. Although the word "image" is commonly used, it is extraordinarily complex in meaning. Therefore, this analysis refers to the definition of "imagination" to understand this multifaceted signifier better. As per this definition, imagination is the faculty or action of forming new ideas, images, or concepts of external realities that are not present to the senses. Therefore, imagination is considered an idea that cannot be characterized as knowledge. Meaning would not be created in the moment of execution but in visual narratives' formal result, which concerns our consciousness in time and space as reality.

Therefore, while each image makes a new circumstance visible, it would be possible to interpret or (re)present reality in an abstract manner. Furthermore, these images would also be responsible for relating the elements of symbolic association to the collective imaginary. Thus, the collective imaginary, which concerns the discernment between illusion and reality, is essential. In this way, based on the principle of a realistic perfection of the image and a possible collective illusion, this understanding is conditioned by the influence of cultural production in the configuration of the contemporary image in large cities as a physical and digital environment related to the social, political, and cultural contexts.

Consequently, the material and "real" aspects imply the sociopolitical and economic contexts based on the relations in consumer society. Therefore, this mercantile scenario of achievements, appearances, and values exchange refers to the everyday aesthetics and social and urban challenges. Hence, considering the development of Western societies and knowledge of the practical realities, a transcendental reflection would be unable to comprehend what a realistic society entails. Through visual and aesthetic analysis, which is appropriate for the cultural system and productive relations, a critique is built about the image context of the political-cultural effects of daily life in metropolises. On the one hand, the digital presence transforms the environment and depicts new ways in which individuals relate to cultures, breaking boundaries of perception conditioned by the understanding and comprehension of habits, beliefs, and, above all, knowledge. On the other hand, the digital presence presents individuals with the possibility of transferring environments and objects through virtual and augmented realities in their daily activities, offering new visual experiences. Through the incorporation of images (static or moving), sounds, and texts in the same digital media—hypermedia, carrying distinct contents—visual contexts with formal diversities have generally exerted significant influence on everyday social interactions, in short on culture and its diversity. Thus, in contemporary times, social media users' behavior has been widely criticized. Among social media users, one type that has been causing controversy and thus becoming the most popular is "influencers." These individuals build their image for the other, that is, for virtual society. Human relationships in social life depend on the rules of the social game. Moreover, the emotions and escapes from said rules can be observed in visual language form. Somehow, influencers are always attempting to expand the field of recognition and popularity. It is similar to artists with their art and the artistic and cultural movements that, to be recognized, need to be endowed with popularity to translate the sociocultural context. Nevertheless, art is understood as alienating individuals from the social system to perceive other realities. Therefore, as is apparent, influencers' social media pages are just one form of communication evolving at the same rapid pace as the emergence of new technologies. Their great strength lies in the possibilities of creating, producing, and being heard and seen. It is the mainstream of our times. Through the exertion of their influence, their behavior is reproduced by people of various types, styles, habits, and social classes. This kind of identification with influencers' behavior motivates their followers to maintain an audience for the continuity of the success already achieved and to imitate them in various pursuits.

Recall Walter Benjamin, who reflected on the role of technical reproducibility in media culture in its evolutionary process in the 1930s. He referred to the success of Walt Disney's character and alluded to the modernism and rhythm of the city of Moscow in his essay "Experience and Poverty" (1933). In that essay on contemporary individuals, Benjamin stated that every individual would like to be Mickey Mouse at Disney World. If one closely examines people's behaviors on social media, especially those of influencers, it seems that Benjamin was conjecturing about the effects of this trend in contemporary times.[1] These characteristics of Mickey Mouse, related to the audience and success, present many similarities with those of stars and icons who change behavior because they involve ideals and enchantment as entertainment without necessarily having any political engagement, but exert significant influence. In "Experience and Poverty," Benjamin (1933) suggested that the image of Mickey Mouse was a dreamed solution to the problems of everyday life. In his view, with the poverty of experience, one should not imagine that humans aspire to new experiences, as they aspire to a world in which they can only flaunt their external and internal poverty. This experience that Benjamin refers to is that acquired with the historical legacy, transmitting cultural heritage, thus alerting us to the lack—to the poverty—of this experience and of knowledge in the face of immediate and superficial experiences. This is just as it is perceived in the sociocultural dynamics today. The contemporary individual is a kind of Mickey Mouse. Mickey Mouse's superhuman abilities are the means and conditions of survival. According to Benjamin, Mickey Mouse's lifestyle fits any historical moment, containing the entire past and the virtual realization of history's final, utopian goal. In his text, the space of modernity is compared with the animated room of Mickey Mouse. He is an inhabitant of modern times, seen as a subject in the process of alienation, subjected to the conditions necessary for his survival and civilization itself. Mickey Mouse has, thus, become the personification of many individuals in modern times, surviving through all of the similarities created based on humans in their vanities, need for recognition, and desire for an audience and followers, shaking the whole of society in its traditional ways of behavior and relationships.

Mickey Mouse is considered a trendy and successful character of modernity, created in 1928 by Walt Disney, during the 1929 crisis, as a model of courage and self-centeredness for the individual of modern times, with a touch of human value. Mickey Mouse has become, for many, one of the most significant symbols in Western history for his commercial success, which is mainly due to the new technologies of communication, moving image production, and cinematography. Disney's popularity and Mickey Mouse's worldwide recognition have surpassed any expectation of a cultural sign. Mickey Mouse was born under the name Mortimer on May 15, 1928. Although only a few lines describe Mickey Mouse in Benjamin's texts, a deliberate image of the successful inhabitant of modernity was sketched. In this vein, many other characters were and still are created, affirming that the existence of Mickey Mouse is one of those dreams of contemporary human beings. It is an existence full of miracles that surpass technical advantages and mock them. The most extraordinary thing about them is that they all, without any improvisation, refer to Mickey. This excerpt, based on Benjamin's writings in 1933, was perhaps the best prognosis of our contemporaneity.

From the pace of modern times to the present day, we are in the third stage of the Industrial Revolution and on the road to the fourth revolution, in a society based on market development, between crises and overcoming, molded on the recipes of success as guidance. In the cultural scene, successful characters present themselves optimistically, emerging from the dream world as a promise of reality, stimulated by the spectacular image of the superstar, and especially with the personification of Mickey Mouse, like Andy Warhol, who Disney invited to link the iconic character with the characteristics of his art in the *Myth* series (1981), which was bought for US$3.4 million in 2011. Since modernity, especially in the universe of art, the identification of American artists associated with Mickey Mouse's success has been common, as Stephen Short (2020) wrote in his article "Things You Didn't Know About Walt Disney's Relationship With Art," besides presenting Roy Lichtenstein's opinion about Mickey, "He's such an American symbol," and adding Warhol's preference—"Mickey Mouse is my favorite actor! Minnie Mouse is my favorite actress! My own favorite personal hero is Walt Disney." Furthermore, Charlie Chaplin, in *City Lights* (1931), was accompanied by Mickey Mouse, and artist Diego Rivera advocated what history has recorded as follows: "Mickey Mouse was one of the genuine heroes of American art in the first half of the 20th century." Moreover, George Grosz stated that Mickey Mouse would be "art in every sense." Damien Hirst is also part of this select group of artists, who said the following: "Mickey Mouse is such a universal and powerful icon." Hirst was also invited by Disney to link his art to the symbolic value of the mouse, with the work *Mickey* in 2014, bought for US$1.5 million.

Not everyone achieves this success in all segments of society and culture—no quantitative research is required to know this. Nonetheless, it is evident that only a few people have this privilege, and this situation can be perceived qualitatively in everyday life throughout the history of their lives. Even though that is a reality for only a few people, many get involved with these images of success. It is a world of enchantment and fortune on one side and disenchantment and poverty on the other, which is evidence of the inequalities and injustices of societies in their visual narratives. However, it seems that the dream will always be present, and maybe this is why these images of the dream world remain and the world is enchanted, even though their lives are in an entirely different reality, one full of disenchantment. Thus, the art of enchantment, making people dream, would be significant for many social causes because it would motivate them to overcome numerous challenges. Mickey's image represents popularization as success and the very model of achievement and accomplishment. However, as an image, it is expressed in a mimetic process by social actors in the most varied niches of society. However, there is no lack of criticism about these sociocultural effects; on the one hand, celebrities identify themselves with Mickey Mouse's success and identity, while on the other hand, others also seek recognition for their art by rejecting not the cartoon but the symbol of American culture that it embodies. In the early 2000s, an exhibition on Mickey Mouse at the San Francisco Museum of Modern Art by Austrian artist Gottfried Helnwein drew attention (Crawford 2006). Unlike all of the Mickey Mouse images ever produced, Helnwein presented him with another face of personality, another expression, intending to provoke—make visible—the profit goals of the Disney Company through this visual criticism and humorously so. The humor in Helnwein's artwork is significant due to the intentionality of humorous practice that always implies criticism of social behavior. In this case, it is not specifically a criticism of the character itself, that is, the charismatic cartoon of a mouse. Rather, it is a criticism of the symbolic value that this character conveys, which is that of American culture. The American corporate institution keeps him alive, and therefore, Mickey Mouse is the strongest icon of this culture because he represents it beyond even the very institution that created him. Mickey Mouse has surpassed his creator and stands beyond and solely as a symbol of the "American way of life" in all its cultural, social, economic, and political structures.

The association with Mickey Mouse's identity refers to the identities of individuals that are manifested and shaped by the institutions of consumer society. On the one hand, the artists who associate their images with that of Mickey ally themselves with all of the symbolic value of the thriving American culture. On the other hand, by humorously criticizing Mickey's image, these artists also criticize North American culture and the real influence that it exerts on the global market. In this sense, in terms of Freudian studies, it is understood that institutions are subject to jokes because they represent authority. Thus, through humorous criticism, the intention is liberation from the pressure exerted by consumer society (Crawford 2006). The association of Mickey Mouse's image by celebrities today only reinforces a status shared by both his successful image and those of stars. Besides the appropriation of Mickey's image in their art as a resource of relevance, this art is also guaranteed to always be present in the collective memory of generations. The generation that saw Mickey Mouse born, knew him, and grew up with him saw that he was also a character animated with humor and a sense of rebellion against the authority of the time. Mickey Mouse, in the 1930s, was a caricature of the modern individual who, with a little luck, could survive the Great Depression. Mickey, with a few resources, could buy a house and a car and have a girlfriend (Minnie) and a pet (Pluto). However, above all, he had the chance to use technological advances and progress. This is the allusion that Benjamin offered us, that of a character who, at the pace of the development of industry and technology, had the pleasure of modern and urban life, passively and unaware of his alienation from the market system. However, the 1930s generation, Mickey Mouse himself, and the whole of society grew and changed and, from the simple experience of life at Walt Disney World, began to experience life beyond the borders on the World Wide Web, under the hegemony of American culture. If, on the one hand, the continuity of the economic hegemony of the United States is currently being questioned, then on the other hand the influence of North American culture is still being considered, which continues to have effects on global culture. Anyway, this will be a new story. What is essential, however, is not merely some resemblance to the individual of today becoming popular, but rather the exact condition is of that individual being manipulated as Mickey Mouse was by the cultural industry. Just as Walt Disney sketched and designed Mickey's every movement so that he could come to life, the same is done to each of us by the designers of social media, of online platforms, and of all digital applications and systems. We are manipulated, controlled in our attitudes and behaviors, and persuaded. Nevertheless, real-life experience and knowledge seem increasingly necessary if we are not alienated in every way. From this perspective, a recent film revealed how digital media can control us, namely *The Social Dilemma* directed by Jeff Orlowski (2020).

The film is a documentary with special participation from Tristan Harris, who worked as an ethical designer for Google and cofounded the Center for Humane Technology. Harris demonstrated the "tech industry problem" by questioning whether we are all bewitched. In addition, *The Social Dilemma* features stories from leaders and experts, such as Jaron Lanier of Virtual Reality Computer Scientist; Roger McNamee, one of Facebook's first investors; Justin Rosenstein, who was an engineer at Facebook, Google, and Asana; and Jeff Seibert at Twitter. The film also features others who have developed technologies associated with persuasion techniques scientifically grounded in neurology and psychology to control users of Facebook, Instagram, YouTube, Twitter, and Google, with the following sole purpose: domination and capital accumulation. Why were they reporting all of this? The answer is simple: They have abandoned their work related to the misuse of technology in favor of an essentially human cause, and they believe that it is still possible to avoid worse consequences for society. Technologies alone do not cause harm to human beings, but how those people make use of them can cause significant damage to humanity.

Extending the Public Image on the Web

Transformations are considered the main feature of urban space as representations of a conception of the world with its characteristics and needs, which vary according to culture, society, and the economy. In its current transformation process, the urban image is the reality of society through the diversity of visual configurations. Furthermore, verbal or written languages are also considered visual representations by remitting the thoughts formed by images. Thus, the action of the present develops around images of the economy, techniques, science, ideologies, and politics among others. These actions transform evolution, and from such a perspective, it is possible to reflect on the conditions that make innovations possible and the construction of everyday facts. Among them, it is possible to mention the configuration of images that characterizes the exception in general rules, transforming the predictable, and everything is viewed as either conventional or traditional. In general, the positive aspects of digital culture are precisely in the belief of an equal relationship between different individuals and the democratization of the Internet. However, among the many communities formed on the World Wide Web—blogs, social media platforms, and video channels—what is perceived is not democratization. It is a fictitious equality, a constant unreality among different people relating to equality only because they create avatars, profiles, and characters that virtually live out their idealizations to resemble or arouse interest in others. These characters come to represent unrealized desires in the real world, a virtual world sublimating the real world. It can even invert everyone's lives' most complex values and conditions. It is life simulated and dissimulated.

The widespread use of selfies on social media leaves no doubt about this reality of posing and moments of success and the perfect life. Anyone can make a social media account and produce images, especially their own images, having at their disposal refined and easy-to-use resources for applying filters, which manipulate images into conditions idealized by the user. The experience of sharing these ephemeral moments—mainly artificially used or entirely fictitious—stimulates not only senses of reward, pleasure, and happiness that are not true but also specifically competitiveness among other users who share the same practice. The dimensions of this social media reality are immeasurable in terms of what concerns the psychological and ethical conditions of human behavior, seeking affirmation and attention, and, above all, sublimating their most varied desires. More recent studies have discussed the emergence of new ethical dilemmas, or whether existing dilemmas are similar to dilemmas in other fields of research despite technological influence, particularly concerning fundamental rights and freedom, bringing the context of privacy into the discussion (Nissenbaum 2009; Buchanan and Zimmer 2018). Building on discussions of the complexity of understanding regarding privacy, Nissenbaum focused on why advances in information and communication technologies provoke uncertainty, protests, and resistance in the name of privacy. Thinking on the Internet, in a context of multifaceted scenarios that imply the user's exposure, is also to understand that positions on the private–public relationship are conditioned various situations. Some explanations derive from the relationship of privacy with anonymity, secrecy, confidentiality, and public versus private life in its most intrinsic aspects, leading privacy to solitude. On the one hand, besides the constitutional rights of privacy, a need exists for personal information to be understood as a fundamental right. Yet, on the other hand, there is a form of improving sociability, fighting loneliness, seeking new experiences among people, and considering many of the fundamental and humanitarian causes of inclusion, based on mutual trust, on solidarity for the best possible relationships among human beings.[2] However, threats of surveillance systems in public spaces are also a reality, and monitoring actions in online transactions and social media behaviors, as Figure 1.1 shows. As recently as 2009, when Nissenbaum wrote her book, *Privacy in Context: Technology, Policy, and the Integrity of Social Life*, the world had not yet experienced the Cambridge Analytica case. However, then, with the movements against social media, precisely because of this case involving Facebook and Cambridge Analytica, the other social media platforms owned by Mark Zuckerberg's company lost credibility and many followers. Presently, many others remain. In general, the reasons for the exposure of individuals on the Internet are numerous and complex. How they are exposed depends not only on the systems, programs,

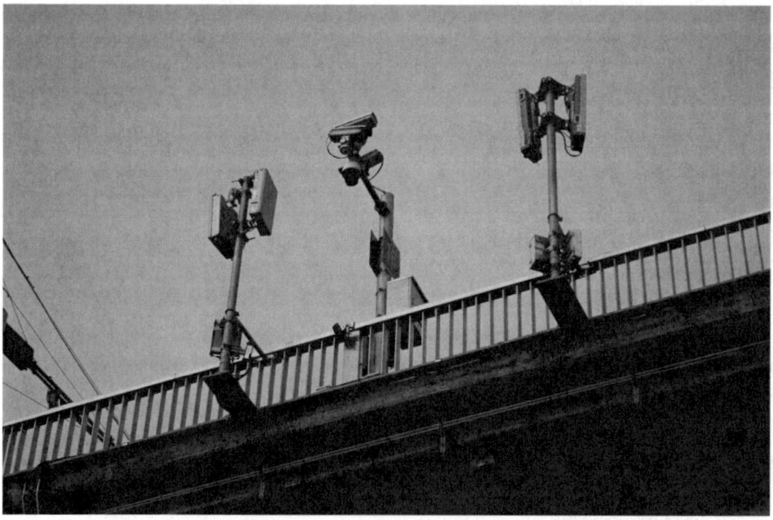

FIGURE 1.1 A city under surveillance.

types of media, and interface design but also on how to use them in terms of their technical, ethical, and moral aspects. Moreover, such use involves social life in general, not only on digital platforms but also in public spaces under surveillance, as Figure 1.2 shows.

Society must be considered in all of its dimensions. The Internet is just an extension of it. When looking at culture and the image value as a vehicle of progress or innovation, everything is focused on the spectacular. Guy Debord (1967) discussed some significant thoughts on this issue in *La société du spectacle* (*Society of the Spectacle*). More contemporary discussions by Jean Baudrillard (1974), who analyzed the consumer society, are precisely related to the context of the public image extended to the Web vice versa. Baudrillard demonstrated the environment of consumption and the consequences of urban life subtracting from several occasions' fundamental existential values. Examples include attention from friends or relatives, the approximation between people, and the increasingly accelerated rhythm due to commercial commitments becoming increasingly complex. In this urban rhythm, some existential values are lacking, which are replaced by dreamed experiences—sublimation—in the public environment, cafes, bars, restaurants, shopping centers, public parks, and especially social networks on the World Wide Web. The last one is a constituent of possible dreams, in short of the meaning of social relations due to industrial development and technical progress. No matter what metaphor is applied, the entire image is consumer society and urban dynamics—the image of individuals alienated or conditioned, with their pleasures and behaviors, converted into actions that maintain the system in its operation and continuity. Thus, since its emergence, humans have followed the Web toward the future as an extension of the public space and where urban art manifests in its various forms.

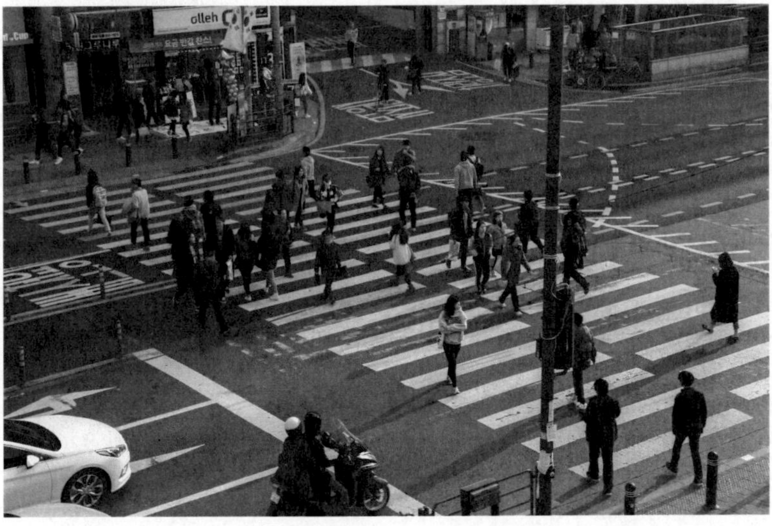

FIGURE 1.2 Public life under surveillance.

Deterritorialized Spaces

Through the arts during the Cold War, especially science fiction literature and movies, the world read and saw fiction stories about incredible secret missions being undertaken using high technology; an example is the film *Wargames* (1983) directed by John Bradham. The film is a fictional narrative about a student hacker who unwittingly connects to and launches an American military supercomputer programmed to execute a nuclear war against the Soviet Union. Subsequently, these fiction stories concerning the digital world in all its revolutionary ways became a reality and represent a change, for better or for worse. The first computers achieved their best market performance after being used by the state for defense strategies. This scenario influenced many children whose parents represented the counterculture of the 1960s and today are part of a generation that became critical to the largest of all revolutions of the 1980s: The Technological Revolution. In addition, an outstanding commitment to what became known as sustainable development emerged during that time. More recently, in 2011, this became the responsibility of a high-tech and deterritorialized youth who, struggling for democracy, used new technologies to achieve one of the most significant events of recent times through the Internet with a "non-profit organization advancing human rights and freedoms by creating and deploying free and open-source anonymity and privacy technologies" (The Tor Project).[3] These youth achieved what could not be imagined at that time, namely a revolution resulting from the search for democracy in the Arab world. Therefore, the question that arises now, in another stage, is how to organize the power in these countries, which are living through a troubled period of transition. Even more recently, the world has been witnessing a global communication war over Russia's invasion of Ukraine through social networks and the media.

Thus, it is necessary to understand the need for technology, science, and rational control; to review some concepts theoretically; and to relate them to arts and communication. Jürgen Habermas situated communication with the consequences of rationality, namely strategies, the use of technologies, and the appropriate organization of systems, as the conditions imposed in a rational society by focusing on the values of the Enlightenment in terms of Max Horkheimer and Theodor Adorno's concept of "instrumental reason" (*Instrumentelle Vernunft*). Adorno and Horkheimer, in *Dialectic of Enlightenment* (1947), distinguished two meanings of the faculty of reason, the first in its relation to absolute truth and the second in its relation to knowledge. For both relations, reason is ambiguous and dialectical. On the one hand, it liberates human beings from subservience, bringing clarification. On the other hand, it conditions them to technocratic consciousness in the service of the development of capitalism and the economic interests of the dominant class. Weber's concepts (1922) of value-rational and instrumentally rational actions (*Zweck-Mittel-Rationalität*), which refer to the technical-rational appropriateness of the means to achieve an arbitrarily elected goal, are also relevant. The concept of communicative rationality was developed in Habermas' work *The Theory of Communicative Action* (1981). Therefore, we consider the need for actions that control this system, that of nature or society, or a way to obtain control. However, this does not mean that this control is exercised as a political act but rather that it is contained within politics. We understand, therefore, that the sense of a political existence is intrinsic to the individual in society, in their public sphere, as a citizen. With technological transformations and the digital era, the whole relationship established between the productive forces, which have increased and become institutionalized by this technological and scientific progress, has exceeded the proportions identified by the theorists who lived under the conditions of a repressive apparatus. This repressive system would produce articulated divisions, whether they were constructive or destructive social work, that conditioned individuals living under the state's repression. This repression or lack of autonomy or freedom of speech would lead to discontent. It would be conducive to revolution even if the individuals lived with a sense of powerlessness under a repressive force. Thus, what are the limits of freedom of speech? What is the sense of freedom for the arts in public space?

Liberal democracy made the Technological Revolution possible, while the advent of globalization and the Internet made communication without borders possible. Through Habermas' answers to Herbert Marcuse's analysis in *Technology and Science as Ideology* (1968), one can partially think of appropriate responses to the most critical questions concerning the interdependent relations of the current system. For this thinker, one can observe a paradox in the understanding of repression. Marcuse believed that repression would disappear from the individual's consciousness because the legitimization of domination produces a new situation resulting from that domination's natural advantages and products, which always offer greater benefits. Marcuse also expected an equivalent liberation for the individual concerning the necessity and the sacrifice of energy and time for technological development. In industrial society, modern and democratic culture has been developed through the dynamics of mass consumption of cultural goods that renounced the forms that the art of antiquity offered, which were the means of transferring desires and their realization through art. In psychoanalysis, this process is called "sublimation." Instead of finding ways to sublimate desires through art, in modern society pleasure is found through material and concrete resources—a material culture, as opposed to an ideological culture. Therefore, it is a less affirmative culture that characterizes modern culture, which has not established mechanisms to transform reality gradually. Still, in this period, without sufficient time for sociopolitical changes to follow the necessary trajectory required for new ways of interpreting the developing system to emerge, these theories were intensively discussed and valued in Latin America, especially in Brazil. In 1964, following the *coup d'état*, Brazil entered a period of military rule that lasted until 1985. This period ushered in an extremely antidemocratic and repressive scenario, where the army comprehensively exercised its power. The theories of Adorno and Marcuse were appropriate to the reality of this period. They supported the first research into sociology, communication, and culture and were exceptionally resistant to the manipulation of the state. On the international scene, Adorno's criticism was scarcely discussed. Thus, before he died in 1969, Adorno understood that his theory of modernity had confronted the decline of modern art; kitsch was assuming more space with the established powers of the cultural industry, who were subject to the market's profits. Hence, today new technologies prevail for production, dissemination, and artistic and cultural practices that do not discredit the theory of Adorno and Horkheimer. They had foreseen the risk of a standardized and globalized technoculture that would be detrimental for the individual in terms of their unique and diverse experiences.

In any case, the existing relations of production are presented as the forms of organization necessary to a technically rationalized society. This rationalization, elucidated by Habermas, demonstrates the ambiguity of Weber's orientation (1922). On the one hand, to present criticism of the development of the productive forces as a measure of discontent, which was meaningless and part of the historically outdated relations of production, allows for the interruption of repression. On the other hand, they can be regarded as measures for justifying the relations of production as an institutional status that is appropriate to the fundamental objectives. In short, Habermas also demonstrated that Marcuse did not know about the consequences of arbitrary scientific and technological development, which today have resulted in a configuration in which productive forces experience new interactions with the relationship of reproduction. Indeed, the writings of Marcuse were mainly derived from the 1950s and 1960s. Today, we must think not only in terms of a quest for demystification but also for political enlightenment (*Aufklärung*) and legitimacy. Understandably, collusion occurred between the state and society concerning the advantages that all of this scientific and technological development offered. On the one hand, the power of the state is exercised in search of control, while on the other hand, resistance in search of the rights of freedom is concentrated as a mass power. Therefore, two opposing forces of equal capacity that establish a point at which the differences are neutralized are a balanced system for the conditions of their own survival.

It seems likely that the personal, professional, and social implications of the Technological Revolution have changed individuals' behaviors. In the face of new digital demands, for example, ubiquity is required, that is, the mobility (or not) of the individual, with the possibility of moving their image, or voice, to other environments being virtual in presence. Consequently, these technological possibilities have made the individual's presence much more viable. It is the simultaneity of supply and demand, action and reaction, cause and effect, in different locations. It is the cause and effect of the image in its understanding, as the medium of inventions and artifices, in an environment not incremented but modified by the possibilities of visual intervention. A transformation of relationships is perceived, which gradually interferes in several areas. First, the social media phenomenon has enormous potential because it allows people with little technical knowledge to have a personalized network account and to post textual and visual messages to millions of others, thus "democratizing" or popularizing to the extreme the mass communication process. Second, social media, as image and spectacle, has artistic characteristics. Because of the easy-to-use technology, social media platforms are channels for free expression, but they must respect the ethical and moral values that prevail in different cultures. Third, they allow communication contrary to the static print form and the technical and cost limitations of the electronic media of radio and TV, allowing interventions, continuity, and ramification.

Furthermore, the arts, free speech, and debate through social media allow users the right to articulate in the face of public exposure. However, when the intent focuses on the power to influence the public, some economically impoverished classes—especially those without access to the Internet—are excluded from this participation supposedly understood as democratized space. Meanwhile, momentary democratization occurs when the individual Internet user participates. Consequently, the economically excluded classes, without access to the Internet, end up consuming the ideas and desires of the upper classes, even though they are, daily, in opposite living conditions. The privileged classes are called "opinion makers" because they have access to all information in the first place, consequently transmitting to the lower classes their opinions through other media.

The facts themselves and stratified images form the cultural image, contributing to "choices" of what was already chosen, from the dream already dreamed toward realizations of what already exists. Information is understood through digital technology as the leading actor in the image of culture, in the interdependent relations between arts, politics, society, economy, science, and technology, under interference from opinions of the favored classes, not only from a supposed controlling state. Depending on the circumstances, whether in a democratic state or not, the selective forms of information provided by the media and digital media must be considered. This means that choice can either be freely acquired or passively influenced by opinion makers. However, in the first instance, the "truth" is not related to information but instead to knowledge, grounded in education above any ideological value. Therefore, this discussion related to "Cultural Images: Real and Imagined Experiences" leads to the understanding that information is guided by the practical reason of opinion-forming throughout media, determining sociocultural and political contexts. Consequently, it is reflected in the arts' functions to educate, inform, entertain, and stimulate perception, sensitivity, cognition, expression, and creativity, and finally in its social function.

Based on the arguments presented thus far, the aim of this discussion is to contemporarily define the effects of images on the formation of realities. Therefore, for the cultural image phenomenon and the complexity of human perception faculties and social imagery concerning aesthetic, political, and social aspects, the arguments have demonstrated the potential of the image as a true power of effects. This was evidenced through a qualitative inductive analysis based on concepts, theories, and research in arts and human and social sciences, in response to the questions presented regarding the ephemeral image in constant reconfiguration by digital media and interactive social networks. The focus is on immediate experience to the detriment of intensive and individual experiences, transforming aesthetic and cultural values. Also discussed was the possibility that these are just reconfigurations into new forms of representations of already established cultural values, besides considering that art, image, and media spectacle construct experiences without even the need to live them. Individuals' identification with the characters of this media spectacle, especially celebrities, offers a dream experience that encompasses the imagination. This imagined experience pertains to the human capacity as the main guidance responsible for its evolution and transformation, in which the imaginative process can (re)create lives in the world or worlds in the individual's life under the effect of images with an ideal of perfection, in an illusory and constant search for satisfaction. It is these types of experiences that were discussed by alluding to the personification of Mickey Mouse, based on Benjamin's essay "Experience and Poverty" (1933), associating it with the success of artists and influencers, all of whom influence the audience. What is fundamental in the sociocultural context of our contemporaneity are the structures of institutions and the capital invested in advertising and strategies for commercial maintenance. Capital is, therefore, the primary mediator between the real and the imaginary, in which (re)creation is a powerful source of persuasion.

What is seen in the World Wide Web is the high potential of widespread interference and power to become an essential tool for sustaining the strategies of the social game. It is mainly a device for artistic or political promotion through which the individual, in transformation with these techniques and language, obtains symbolic value. Regarding Habermas' conceptions, it is understood that technological culture follows a political orientation to shape the cultural image and present the purposes stated by Weber, that a rationalized development would be a rationalization of social structures, thereby also becoming the real reason for the development—that of maintaining the references that respond to the fundamental interests of the state and the community. According to arguments developed and disseminated by the media, citizens' actions hold private and public spaces as well as cultural image configurations. The images associated with relevant issues determine the public's responses and empower public opinion. Thus, from these positions on the image of culture, a grounded consideration becomes necessary when dealing with images that portray social reality related to the human capacity to create narratives that constitute a collective consciousness. Therefore, next in "Urban Space and Collective Consciousness," the main objective is to analyze the influence of visual narratives on social reality. It is known that contemporary reality involves society mainly through the effects of images disseminated by the media. To realize that the dominance of these images prevails in defiance of optimistic stimuli and goodwill also means recognizing that the notion of reality has lost its meaning and its absence persists.

Notes

1 Marc Jimenez (2005, 2011), a French philosopher and German studies scholar specialized in aesthetics from Université Paris 1 Panthéon Sorbonne, indicated this analogy of the contemporary individual with Mickey Mouse.

2 The following excerpt from Helen Nissenbaum indicates other positive aspects as well:

> "To be sure, not all systems that alter the flow of information are cause for alarm, for there are clear cases of new information devices and systems that serve societal as well as context-based values, ends, and purposes better than those we already have in place (e.g., promoting intellectual development, health and well-being, and vibrant democracy)." *Privacy in Context: Technology, Policy, and the Integrity of Social Life.*
>
> (Stanford, CA: Stanford University Press, 2009), 3

3 The Tor Project's history: "The Tor Project, Inc, became a 501(c)(3) nonprofit in 2006, but the idea of 'onion routing' began in the mid 1990s," The Tor Project, accessed September 20, 2022. https://www.torproject.org/about/history/.

Bibliography

Adorno, Theodor. *Ästhetische Theorie* [Aesthetic Theory]. Frankfurt am Main: Suhrkamp, 1970.

Althusser, Louis. "Idéologie et appareils idéologiques d'État (Notes pour une recherche)" [Ideology and ideological state apparatus (Notes for a search)]. *La Pensée*, no. 151 (1970). Reprint, in *Positions*, 1964–1975. Paris: Les Éditions Sociales, 1976.

Althusser, Louis. *Sur la Reproduction* [On Reproduction]. Paris: PUF, 1995.

Baudrillard, Jean. *Simulacra and Simulation*. Trans. by Sheila F. Glaser. Michigan: University of Michigan Press, 1994. Originally published as *Simulacres et Simulation* (Paris: Galilée, 1981).

Baudrillard, Jean. *La société de consommation: ses mythes, ses structures* [The Consumer Society: Myths and Structures]. Paris: Gallimard, 1997. Reprint, London: Sage, 1999.

Benjamin, Walter. "Experience and Poverty." In *Die Welt im Wort* (Prague). Gesammelte Schriften, II. Trans. by Rodney Livingstone (December 1933): 2013–2019. Reprint, *Walter Benjamin: Selected Writings 1927–1934*, edited by Michael W. Jennings, Howard Eiland, and Gary Smith, trans. by Rodney Livingstone and others, 731–736. Cambridge, MA: The Belknap Press of Harvard University Press, 1999.

Benjamin, Walter. "Das Kunstwerk im Zeitalter seiner technischen Reproduzierbarkeit" [The Work of Art in the Age of Mechanical Reproduction]. 1939. Reprint, *Medienästhetische Schriften*. Frankfurt am Main: Suhrkamp, 2002.

Buchanan, Elizabeth A., and Michael Zimmer. "Internet Research Ethics." *The Stanford Encyclopedia of Philosophy* (2018). Edward N. Zalta, ed. https://plato.stanford.edu/archives/win2018/entries/ethics-internet-research/

Cadwalladr, Carole, and Emma Graham-Harrison. "Revealed: 50 Million Facebook Profiles Harvested for Cambridge Analytica in Major Data Breach." *The Guardian*, March 17, 2018. https://www.theguardian.com/news/2018/mar/17/cambridge-analytica-facebook-influence-us-election

Crawford, Holly. Attached to the Mouse. Disney and Contemporary Art. Maryland: University Press of America, 2006.

Debord, Guy. *La société du spectacle* [Society of the Spectacle]. Paris: Gallimard, 1967.

Habermas, Jürgen. *The Structural Transformation of the Public Sphere: An Inquiry into a Category of Bourgeois Society*. Trans. by Thomas Burger with the assitance of Frederick Lawrence. Cambridge, MA: MIT Press, 1989. Originally published as *Strukturwandel der Öffentlichkeit. Untersuchungen zu einer Kategorie der bürgerlichen Gesellschaft* (München: Luchterhand, 1962).

Habermas, Jürgen. *Technik und Wissenschaft als Ideologie* [Technology and Science as Ideology]. Frankfurt am Main: Suhrkamp, 1968.

Habermas, Jürgen. *Theory of Communicative Action*. Trans. by Thomas McCarthy. Boston, MA: Beacon Press, 1984. Originally published as *Theorie des kommunikativen Handelns* (Frankfurt am Main: Suhrkamp, 1981).

Hegel, Georg Wilhelm Friedrich. *Lectures on the Philosophy of Art*. The Hotho Transcript of the 1823 Berlin Lectures. Trans. by R.F. Brown. Oxford: Clarendon Press, 2014. Originally published as *Vorlesung über die Philosophie der Kunst*. Berlin 1823. Rewritten by Heinrich Gustav Hotho (Hamburg: Felix Meiner, 1998).

Horkheimer, Max, and Theodor Adorno. *Dialektik der Aufklärung* [Dialectic of Enlightenment]. Amsterdam: Querido, 1947. Reprint, Frankfurt am Main: Fischer, 1998.

Jennings, Michael W., Howard Eiland, and Gary Smith, eds. *Walter Benjamin: Selected Writings 1927–1934*. Trans. by Rodney Livingstone and others, v. 2. Cambridge, MA: The Belknap Press of Harvard University Press, 1999.

Jimenez, Marc. *La querelle de l'art contemporain* [The Quarrel of the Contemporary Art]. Paris: Gallimard, 2005.

Jimenez, Marc. "Walter Benjamin." *Revue Mag Philo*, January 2011. http://www.cndp.fr/magphilo/index.php?id=95

McLuhan, Marshall. *Understanding Media: The Extensions of Man*. New York: McGraw-Hill, 1964.

Nissenbaum, Helen. Privacy in Context: Technology, Policy, and the Integrity of Social Life. Stanford, CA: Stanford University Press, 2009.

Short, Stephen. "Things You Didn't Know about Walt Disney's Relationship with Art." *CoBo Social International Co.* https://www.cobosocial.com/dossiers/walt-disney-with-art/

The Tor Project. "History." Accessed September 20, 2022. https://www.torproject.org/about/history/

Weber, Max. *Wirtschaft und Gesellschaft: Grundriss der verstehenden Soziologie* [Economy and Society: Outline of Understanding Sociology]. Berlin: Peter Kietzmann, 1922. https://www.textlog.de/weber_wirtschaft.html

2

URBAN SPACE AND COLLECTIVE CONSCIOUSNESS

After dealing with the effect of images and the value of exposing different ways of perceiving and imagining realities by discussing the media spectacle in "Cultural Images: Real and Imagined Experiences" (Chapter 1), which concerned individual appearances in the construction of public representation, the focus—in continuity—is now on the configurations of a collective consciousness throughout visual narratives and communication actions. Therefore, the dichotomy of the imaginary worldview and the real world is imbued with all of its positive and negative aspects. The constant creation, the (re)creation, through the arts relates to the ongoing message that maintains the continuous relationship of our existence as a communication process throughout our lives and those of future generations. Many visual productions seem to be created for the public to convey an imaginary universe and therefore represent the totality, such as environments, scenarios, and other fantasy worlds, as well as works of art in the everyday reality of urban centers.

Therefore, it is assumed that the visual manifestations of social issues shape the dynamics of the worldview, structuring the productive forces and their reproductions (Benjamin 1939; Althusser 1970). In this vein, in the daily practice of communication, the image seems to be presented as an essential element. Consequently, the ideas represented in images would be perceived as the power of cultural institutions, which are capable of molding society through economic and political development objectives. Art today, in this sense, is supposed to perform following the ideals of cultural and state institutions, or in a pseudo-consensual way. This art would be provocative to the system when facing freedom of expression restrictions in a counter position (Bobbio 2007). Furthermore, it is necessary to situate this analysis amid aesthetic and critical theories (Adorno 1970; Habermas 1962, 1981) to the present day to consider the current

DOI: 10.4324/9781003314943-3

image's significance by asking the following questions: Would artistic realization manifest in the contemporaneity an interdependence of the institutional cultural structures and sociopolitical relations? What is freedom of expression in the arts? Which personal or collective experiences have a space in the arts within contemporary society? In search of pertinent answers, this arts-based investigation aims to understand the complex worldview through lived experiences from the point of view of social reality, which would be in the visual narrative perspective (Schoolman 2020). Among the research methodologies linked to this narrative character are various perspectives linked to qualitative arts-based research. The scope is artistic manifestation in urban space delimiting the urban experience, which incorporates the investigation of visual narrative forms to connect ideological abstractions with personal and collective situations of cultural experience.

The development of this analysis is divided into three stages. The first, "The Phenomenon of Consciousness Reaching Freedom," considers the concept of freedom based on illusion related to the sense of truth and reality, which would condition social relations in the formation of collective consciousness, through actions in urban space, including artistic manifestations. In the second stage, "The Art Consciousness of Urban Space," artistic and communicative actions are analyzed concerning fundamental rights. In particular, those related to the media and images (i.e., freedom of expression) and associated with maintaining social order through equal power and choice with integrity toward a consensus are analyzed. Finally, the third and final stage, "The Visible Dynamics of Urban Space Toward Democracy," discusses the universality of images in the face of contemporary metropolises through their daily life narratives. In this context, the intrinsic values of the nation (i.e., language, tradition, art, and culture) are exposed in the public sphere, in the national territories, and then globally.

The Phenomenon of Consciousness Reaching Freedom

The notion of freedom is philosophical and through aesthetics it applies to artists' state of being free and to their art. Therefore, freedom is the central concept and illusion is a secondary one for positioning the quarrel of being free, relating it to the dichotomies of truth and reality. Through this problem, it will be confirmed whether art—the object of study in aesthetics related to the pleasure of beauty, and its reverse—would condition human goals or not, mainly through the art world, to understand human evolution regarding the notion of freedom.

According to analytic philosophy, we are conditioned by the vicissitudes of language, tradition, and the history of Western thought, which situate us in a sociocultural system to understand images as a means of communication. Likewise, idealists developed systems and theories to study the phenomena of consciousness—to learn about the apparent manifestations of the spirit—in terms of Hegel's approach to reality as a historical process. In Hegelian philosophy, the spirit is in a historical process, the purpose of which is development and evolution. Thus, this approach of the concept of collective consciousness under

Western, analytical, and Hegelian thoughts will be related in light of some of the observations of Jürgen Habermas on the contemporary context of media development.

In his work *The Structural Transformation of the Public Sphere*, Habermas ([1962] 1990) discussed the assumption that a society is developed by its common interests, through media, the judicial system, and the state's political power, by systematizing public dynamics in its complex grandeur. Habermas demonstrated the formation of a collective consciousness that results from rational agreement within social relations. Actions are performed in the public space, and consensus is sought as an indication of success. Habermas presented a rational notion of social reality. He discussed the process of modern state development in relation to the private sphere when the latter becomes public, and in which public opinion builds the public sphere.

The foundations for this context have their origins in the Enlightenment. The greatest reference in the 18th century to a public and politicized life was the French Revolution, not only for France but also for Germany, as confirmed by Habermas, and of course for the whole of Europe and a significant part of the world. Literature and the arts have come to be conceived as contributions to political discussions or allusions to political events. However, the most significant focus is not on the aspects of urbanization or the ideologies of the generation that marked the cultural revolution, nor on the consequences of economic and social development. It is not even on the characterization of the process of transformation due to the emergence of new policies and commercial relations that guide the market system. Rather, the emphasis is on the analysis of the consolidation, by consensus, of society's orientation and meaning, particularly the influence from the 18th and 19th centuries, when literature became increasingly important to the bourgeoisie, giving meaning to mass culture. Our illusion of reality is due to the influence exerted by art and literature, as well as by technological resources, which build mass culture through the media. Accordingly, the public sphere has been configured by all of these influences. Hence, society is conditioned to adapt to transformed models through a process that sustains itself, without significant risks or considerable controversy. However, the differences would be noticeable if the extremes were compared over time. Moreover, these differences bring changes independently of a desire to innovate. Thus, cultural transformations and models are the results of both decisions and a process of continuity of values through confirmation, that is, the establishment of the consciousness of social reality. The importance of this process is not only its origin but also its implementation, with errors and corrections throughout the experience. Of course, life depends on ethics for aggregation between individuals in social reality. Without them, life could not continue in society, decisions would not be the result of common agreements, and models would not serve the common good. The question of innovation in this context seems unlikely. Humans would find innovation only in the breaking of their relationships within models, resulting in paradigm changes. It is understood that models, as with all presupposed relationships, are preestablished with the intention of concretizing a "standard assumption of relations" so that the public space forms the collective consciousness. However, there is always room for improvement.

Society is continuing to be presented with a growth of media power, while the search for a balance between the freedom of expression and equality of rights remains. Such a significant complexity of needs in a contemporary democracy can also be observed in competitive challenges regarding a common agreement, resulting in a collective consciousness. Today, many institutions, communities, and associations have interests in common with many others. Among the many interests, the most detached are cultural events. For this reason, the state has been subsidizing arts and educational initiatives, such as museum exhibitions, concerts, and festivals, instead of the previous subversive achievements that prevailed in modernity. Moreover, a relationship of exchange exists, meaning a public relationship with common goals and an agreement to attend to the models of a society's interests (Habermas 1981). Habermas's considerations were related to the purpose of rational communication, based on language and discourse while searching for an understanding—a consensus—between differences as a solution to problems in all areas of society, in both their ethical and aesthetic aspects. Opposing these considerations, Jean-François Lyotard (1979) argued in *La condition postmodern* (The postmodern condition) against consensual resolutions, defending the differences in the discourses. Yet, is it not precisely from these differences that a consensus can be found? Therefore, the challenge seems to be the capacity of the majority to develop dialectics, or rational communication in the domain of emotional actions. It is precisely this emotional aspect that predominates many artistic and cultural achievements. This approach follows in the consequences of cultural production.

To contemplate the rational aspects that predominate in Western knowledge toward freedom, let us return to the basis of Cartesian thought. According to *The Principles of Philosophy* of Descartes (1640), consciousness in its operation is conditioned to error, mistake, and illusion. Therefore, the Cartesian doubt is related to the fact that consciousness affirms itself when it finds itself in situations of mistakes and errors; that is, it is always seeking affirmation to deal with everything that is not confirmed to be true. This possibility of truth as theory, based on the thought of Descartes and Spinoza, is understood as an affirmation of freedom through knowledge of the truth (Curley 1985; Nadler 2020). Thus, power can be seen as freedom, which would not be possible without knowledge of the decision, facilitating a consideration of the criteria of truth and error in practical affirmation, especially by technique and science. Moreover, the phenomenon of illusion is a characteristic of consciousness, which, according to Spinoza, is imagination; for consciousness to be free, this would be the consciousness that there is no reality (Garrett 2008).

The truth is the subject that seeks the understanding of this determinism, which would lead to the discovery of a new freedom, namely the understanding of what determines us. According to Spinoza, a necessity of our nature is to know what our essence is and what determines us, whether those causes are internal or external. For him, these are limitations that cause a human being to make mistakes. Consequently, for a consciousness of freedom, it is necessary to understand the conditions of human nature. Thus, truth consists of being free, knowing oneself, and knowing the outer things that are in contact with the truth, with the truth offering the power to be free. It is this freedom that is defined by the understanding of knowledge; it is the relationship between understanding and will. This relation was observed by Descartes as well as represented by Spinoza. This knowledge consists of the understanding of the primary questions. In the thought of Descartes, this means that to be free is to know the truth. This demonstrates to us that free will is an illusion; it is a freedom that is no longer a choice. Humans are encouraged to choose what seems to be the truth and the best; this stimulus that humans have for the truth is due to the inclination of humans for the truth because we would have no other thing. The moment that one views freedom as an illusion, one would be in true freedom; in short, we must know freedom, not truth. The consciousness of the experience of freedom places one in a condition of recognition, then of a consciousness of the possibility of choosing one option or another, and one will wonder about the arbitration and reality of this arbitration; that is, in general, all the difficulty of the subject is a paradox. Thus, we will attempt to overcome the contradiction, and therefore, the paradox. The experience with freedom is an experience of consciousness, and consciousness is an illusory phenomenon by which the fact of choice is consciousness, according to Spinoza. Thus, the fact of choosing, of having the experience of consciousness, is already being free.

Nevertheless, freedom in the arts is one in which the artist is conscious and in full control of the boundaries and rules of creating. It is an achievement within the limits of creation, regardless of the historical context and, of course, the idea of freedom bound to that moment. The artist recognizes the limits that exist in any realization, that is, the problems that seek solutions, thus requiring they/their to always be able to solve them. However, through creativity, it is understood that artistic practice is always linked to the technique and methodology of achievement of the form—the rules of art—which necessitates idealizing freedom. Creations formed without any rules or limits have never been an accomplishment. Somehow, they have always existed to be overcome or transgressed, only for other rules or limits to be imposed—for either subjective or objective reasons—as a manifestation of freedom as well as the autonomy of the artists. Therefore, rules are compatible with freedom, which confirms that freedom is exercised through the power of rules because some edicts arise from a subjective need to confront certain problems or difficulties. Ultimately, the needs to be met require the will to perform. Hegel perceived that the growing assertion of artistic freedom and the autonomy of aesthetics within art resulted in an unlimited autonomy of individual choices. Hegel's aesthetics are part of his philosophy of spirit (Houlgate 2016), which was detailed in his work *Phenomenology of Spirit* (1807). In Hegel's philosophy of art, art comes to be valued according to its characteristics, in particular its form and content, and specific criteria determined by the context, ideas, and place in history, in which art production is an attempt to reflect the social and political reality. Individuals, through Hegelian's reasoning (self-determination), acquire more freedom, meaning they are conscious of their autonomy over their decisions; it is the exercise of one's thoughts, of one's consciousness of life—one's spirit (*Geist*). One's subjectivity finds space to experience itself through language and thoughts while individuals experience their autonomy materializing. In art, this relation to the spirit—the idea—concerns the content of art and its mode of presentation, which is formed materially. Thus, the balance between content and form, that is, a synthesis achieved through a process of dialectical thinking, in which the spirit clarifies the confrontation between thesis and antithesis, is related to a moment in time and place of realization. For Hegel, the work of art reaches its highest level of achievement through the expression of the balance between content and form, technique, and materials, in each epoch and state of consciousness of life, that is, the spirit of the time (*Zeitgeist*). In this realization, the artist's imagination exerts its potential and presents the characteristics of sociohistorical influence and its artistic representations.

Freedom lies in the creation of artwork, not in the sense of liberty as a theme but in terms of the rules within the meaning of autonomy and the means by which artists produce for a determined purpose. Then, two needs for freedom can be perceived: One is subjective, an internal necessity in the sense of liberty idealized in the Platonic sense, while the other is more objective, namely a need imposed by the medium, by the sociocultural context, in the meaning of a poetic freedom influenced by the ideas of Aristotle. Overall, many theories about consciousness exist for explaining the human condition in relation to oneself and one's ambiance. Different ways of thought have been developed regarding this phenomenon and its diversity of features. As Robert Van Gulick (2017) put it, "each in their own way aims respectively to explain the physical, neural, cognitive, functional, representational, and higher-order aspects of consciousness."

The Art Consciousness of Urban Space

One of the most significant modernist milestones, and representative of a new direction for the democratic orientation following the Second World War, was the essential mission of enforcing fundamental human rights through the creation of the United Nations (UN) organization. To that end, a new relationship between institutions and organizations linked to the nation states has been verified for the meaning of contemporary art. This sense of art has slowly shed its subversive characteristics to become a more consensual art associated with influential organizations' political and social ideals. A significant reference in this regard is the importance of the image of artists linking their art to the values of fundamental rights aimed at democratic societies. The main headquarters of the UN maintains a rich collection of artworks preserved by the UN Art Committee. Renowned artists from UN Member States have donated artworks to the collections that are representative of the UN ideals: "Embodied in the Charter of the United Nations: the promotion of the idea of the unity of humankind in all its cultural diversity." Among these artworks are the *War and Peace* murals by Candido Portinari. These two murals were presented to the UN, each of which measures 34 × 46 feet. They were painted in oil on six-sheet cedar plywood and are mounted on the east and west walls of the delegates' lobby on the ground floor of the General Assembly building. The murals were a gift from Brazil to the UN Headquarters. Their presentation was unforgettable, because Portinari could not be present at the inauguration. Ironically, Portinari could not obtain authorization to enter the country from the U.S. government as he was a communist.

Nevertheless, this situation highlighted the meaning of a masterpiece with objectives that overcome differences; that is, art remains a memory and a message for the ideals of the UN. Another remarkable artwork is a mural by the Spanish artist Jose Vela Zanetti, which is also located at the UN Headquarters. The colossal mural is titled *Mankind's Struggle for Lasting Peace*, which depicts destruction and the overcoming of the horror of war to encounter peace. The mural has been described as follows: "Concentration camps, bombings, and all the agony of modern war are symbolized in the painting, in the center of which a gigantic four-armed figure implants the emblem of the United Nations as mankind reconstructs a war-torn world" (United Nations Gifts 1953).

The artist attentive to citizen's demands in the urban space maintains the ideals of democracy in their artistic purposes. Their artworks are situated within the social context, astute to the human condition. This democratic awareness guides the artist to orient themselves toward the search for their rights, in principle, not only for the social ideology but also for the reality of the right to freedom based on other rights, such as the triumph of the citizen's personal sovereignty, which is already established and necessary. However, this is not only a history of the past, as several parts of the planet continue to experience wars and isolated conflicts, such as Ukraine, Syria, Yemen, Ethiopia, Afghanistan, and Iraq. This demonstrates the constant struggle for peace in which humanity finds itself. Even if it is only a part of humanity far from stability, it is necessary to think of society globally, both to fight any principle of war and to maintain peace radically. Otherwise, humanity would be destroying itself.

These rights are related to maintaining the social order by means of equality of power and choice—a shared space for all—with integrity and managed by society itself. Nonetheless, we must also consider that modern democracy does not possess the same sense of democracy as ancient Greece. In fact, modern democracy differs in the diverse systems of government. However, in general, citizens in today's democratic societies assume the existence of equality and seek greater freedom in their decisions regarding what to do concerning their choices, desiring some power and greater participation. Independence established within a community represents the orientation of accomplishing actions without restriction. Accordingly, in social relations, citizens seek greater equality and the right to participate in decisions and decentralize power. The images resulting from this relationship between art and politics focused on everyday aesthetics and democratic ideals are perceived when art's authenticity transforms creative production into a social cause. Thus, we understand the importance of visual studies in contemporary societies amid the accelerated development of visual culture and media. Images from different contexts primarily incorporate the possibilities provided by culture's diverse relationships. Particularly noteworthy here is *Tribute 21* by Robert Rauschenberg, a series of offset lithographs printed at Universal Limited Art Editions (ULAE), New York, which were donated to 21 museums worldwide, including the Museum of Contemporary Art, University of São Paulo in 1994. These images are related to the contemporary impact of the image associated with cultural value as sociopolitical progress, seeking to improve what could be described as a "democratization of art." Another objective of the work is the aesthetic reflection provoked by the images, which involve cultural diversity and respect for difference. There is perhaps no more significant and symbolic example of the visualization of an abstract idea related to human rights than the tribute to Nelson Mandela, reminding us of the apartheid era in South Africa. Other examples from the same *Tribute 21* (1994) series are *Ethnic Cultures—Dalai Lama, Peace—Mikhail Gorbachev, Technology—Bill Gates,* and *Environment—Al Gore.* Overall, the names and themes mentioned in Rauschenberg's *Tribute 21* make the meaning of his art evident, thereby avoiding any ambiguity or interpretive and aesthetic speculation about the visual composition of his groupings, which encompass a global understanding. They represent social facts, and furthermore, the prints celebrate events with humanitarian themes that are interpreted as relevant to this period from modern to contemporary times in their historical, social, political, and cultural significance. In 1995, for the celebration of the 50th anniversary of the United Nations Foundation, an art exhibition was held at its Geneva complex, titled *Dialogues of Peace.* The exhibition featured artworks by 60 international contemporary artists with the concept and curatorship assumed by Adelina von Fürstenberg and organized by The French Association for Artistic Action. Among these artworks, the series by Rauschenberg was conspicuous. However, the exhibition organizer, Fürstenberg, clarified that not all of the works were explicitly political,[1] but "in their different ways, they all make statements about the state of the world half a century after the birth of the United Nations." She also emphasized the immense importance of Rauschenberg's *Tribute 21* series, given that "each of the 21 serigraphs touches one aspect of human existence—from happiness and human rights to health and technology—and, taken together, they cover the range of human experience."

Let us remember the social changes that have occurred since the French Declaration of Human Rights (*La Déclaration des Droits de l'Homme*) in 1789, as a first step, which led to what is stated in the Universal Declaration of Human Rights, approved by the General Assembly of the UN on December 10, 1948, and is still in practice. Moreover, social changes resulting from economic, technical, and industrial development have led to new forms of communication, greater access to information, and acquisition of knowledge. These changes represent a process of social transformation that has led to human relationships acquiring new needs, thus conditioning them to new requirements, which are, in turn, new rights. The diversity and complexity of human rights are predicated on the difficulties inherent to the content of righteousness, which is based on morality and, by extension, common agreement. On the one hand, the consensus of some on a particular subject implies the understanding that these citizens have an "absolute value." On the other hand, the expression "human rights" leads to homogeneity in society before the law; in fact, to give rights to some, others will have had their rights removed (Bobbio 2007, 64). Norberto Bobbio justified the present situation and the future concerning democracy by presenting the most significant right: The right to freedom of expression. However, to obtain the said right, it is also necessary to remove citizens' right not to be deceived, persuaded, provoked, scandalized, or mocked. Therefore, we can imagine freedom and social rights as we can experience them in actual societies, such that the freer humans are, the less factually correct they are and vice versa. Thus, for Bobbio, "freedom" is the right accorded when the state does not intervene, and "power" is the right by which the state's intervention is assumed. Consequently, freedom and power will never be complementary but rather incompatible. Thinking about freedom is part of human nature, that is, of life in all its senses. However, understanding the value of freedom is an essential part of the experience in conjunction with the lack of this freedom in various forms. Humans are conditioned as citizens or simple survivors to the natural conditions that determine their paths. Hence, to think about humankind in its biological, psychological, and social nature; its ethical, aesthetic, and moral aspects; and its cultural, ethnic, and religious differences is to seek an understanding of how humankind, amid determined or arbitrary conditions, can build better relationships. To do so, undoubtedly, communication in all its forms and means is the determining factor. Consequently, regarding communication forms, Habermas presented in *Theory of Communicative Action* ([1981] 1984) a discussion of that assumption on a relationship that aims to reach agreement—a consensus—through reason. Such a relationship is concerned with human rights issues and freedom of expression in an appropriation of the public sphere by individuals through the possible participation and realization of all, as mentioned earlier.

Accordingly, the appropriation of the public sphere means a space built for a public opinion that concerns power; however, it concerns publicity as power and the necessity of visibility, as Immanuel Kant stated. That would be the result of a democratic governance. Bobbio considered this issue of human rights, freedom of expression, and the public sphere, presenting it as a political and moral need. Bobbio demonstrated that freedom is the reason that the public should know. According to Kant, this is what humans can create with the knowledge to generate public freedom and, for him, our consciousness—the fact of reason—is the moral law. The knowledge of the reality of freedom is the practical reality, that is, the conditions of the purpose of experience. In this sense, in a more complex way, we have a scheme for a relationship dynamic based on the desires and purposes of individuals, conditioning them to actions that achieve goals—the categorical imperative—a paradoxical problem of reciprocity between morality and freedom in Kant's work ([1788] 2008). Therefore, for Kant, we know freedom before any understanding or action—*a priori*—in ourselves because morality is also within ourselves. Consequently, we know freedom even before putting it into practice through its relationship to our reason and morality. Hence, in the understanding of Kantian philosophy, the only way to act autonomously is through the moral law in the sense that freedom and practice mutually imply one another.

Moreover, by considering the forms of power that are not transparent and the threat to liberal democracy, we can approach Marcuse and Althusser's ideas and theories. They presented a scenario of the ideological system of the state by articulating the relationships. These are structured on the strength of their power through the means of ideological reproduction. The meaning of these structures is located in the way that governments maintain control and restrict individual freedoms. Therefore, for Bobbio, these theories did not pay enough attention to the importance of democratic polices. The relationship of these rules, whether good or bad, becomes relative when they are evaluated—even more so over time. Here, for example, is the solution proposed by Althusser against the supposed "tactics" of the bourgeoisie to maintain itself in power through reform. Reform as a solution? What does this reform consist of? How do we evaluate all parts of the structure and reorganize them? Which factors are good, and which are bad? Good for whom? Hence, when Althusser ([1970] 1976) presented his interpretation of Marx's work, he was leading us toward a reality that consists of a probable organization of bourgeois power employing strategies that would deceive society to maintain itself in power. Althusser structured relationships between the forms of realization, the actions of the ideological tools of the state's power that, represented by the bourgeoisie, are organized and capable of managing the relationships, and their manifestations through strategies—or, to use his word, "tactics." In all of his work on reproduction relations, according to Marx's work, clarity is essential to understanding the dynamics of this association with values related to individual rights. However, any attempt to convince those of us situated within the democratic governance of the hypothetical reality of the control by the bourgeoisie through employing the reproduction of the state today seems unreal. Thus, the reality is different for those in a totalitarian regime, even in a system under autocracy that does not recognize individual rights.

Nevertheless, for those who have not lived under repression, it is still possible to imagine such a context, the control of the population by a totalitarian regime through the domination of information and communication technologies, through the fictional novel *Nineteen Eighty-Four* by George Orwell (1949), which also exists in film version. Thus, between lived realities and imagined realities, based on sociocritical and aesthetic theories, Bobbio's observation is still relevant, who discussed democracy as involving the need to contain a hidden power within itself. He argued that the comparison between the ideal model of visible power and the reality of things must consider the tendency of all forms of domination to evade the attention of the dominated by hiding and caching, or in other words, through secrecy and dissimulation (Bobbio 2007).

The Visible Dynamics of Urban Space Toward Democracy

The powerful production of the image industry occurs in a favorable context, leading to the universality of images. The lives of contemporary metropolises are narrated and represented through their histories, experiences, anxieties, and hopes, and finally through their daily lives through images that propose projections of the social actor, the individual's behavior, and ideals. This projection of the social actor transmits, for most people, the collective imagination of multicultural space. Moreover, television programs, audiovisuals, film production, and advertising form social imagery behaviors and present images linked to many cultural identities. In this context, the intrinsic values of the nation, namely language, tradition, art, and history, are exposed in the public sphere, in the national territories, and then worldwide. The latter is a production in a globalized culture that consists of a world exchanging fragments of identities. Thus, it is the character of cultural globalization that prompts image mediations and forms new perceptions among the particularities of cultural difference. Therefore, images play the leading role in cultural production, transmitting concepts, ideas, statements, and democratic ideals. However, this global notion of the connection between nations and cultures does not occur freely. State regulatory bodies administer many technical means to control both local relations and access across national borders. This happens not only in countries with autocratic or totalitarian governments but also in countries considered democratic.

The main issue is not the mass industry or how the image is disseminated, locally or globally, nor is it the role of the arts and its audience, but rather all of its cultural effects. An audience differentiated by its form and tastes watches online movie channels, concerts, and theater and visits historical monuments and art and museum collections virtually, even if partially, without the aesthetic experience of the face-to-face moment with the original work, which brings to mind the Benjaminian concept of "aura." Even so, it is also worth noting that most are interested in immediate experiences just for the information they impart. Nor does the issue concern the history of the image in its technical evolution and reproduction, the ambiguous role of the media, and whether they favor the system or large audiences and celebrities. Today, this question no longer has meaning, just as it had no meaning for Walter Benjamin ([1939] 2002) when he revised his 1936 essay "The Work of Art in the Age of Mechanical Reproduction," last published in 1939, after realizing that the decline of "aura" would not result in the democratization of art and culture. Furthermore, in societies that are determined by differences and injustices in their formation, sociopolitical organization, and behaviors, the democratic system itself is placed in doubt. Therefore, nothing can be democratized without an efficient democratic system, especially culture. Thus, *a priori*, the central question is how to make democratic values viable through the arts. In this sense, the main objective of this reflection is to bring aesthetics and politics closer, that is, the correspondence of the images of art with public life regarding the ideals of the organization of a city, state, or nation. Therefore, to think about the object of aesthetic reflection—art related to the pleasure of beauty and its opposite sense—is to think about the reception of art forms and their subjects in contemporaneity interacting with daily social life and its democratic ideals. The consequence of this thinking represents social aesthetics, in which reflection on the arts is linked to political-cultural aspects. Accordingly, among the many ways in which art manifests itself, the experience and effect of a wide perception are most realized by the cinematographic art and, consequently, the moving image.

However, specifically in politics, the focus is on all of the arts and the effects through which they enable politicization, that is, citizen awareness in all its cultural diversity and differences. Regarding this point, the analysis by Morton Schoolman in *A Democratic Enlightenment: The Reconciliation Image, Aesthetic Education, Possible Politics* (2020) is essential. This is because it brings together politics and aesthetics for democratic enlightenment within the process of politicization mainly, but not exclusively, through the art of cinema. Consequently, discussing democracy's essential values also requires one to understand the complexity of its ideal of respect for differences, thus combating violence. In this regard, Schoolman understood that one of the ways to overcome this violence is through the ideal of reconciling identity and difference by means of the "image of reconciliation," thus providing an aesthetic education in relation to democratic enlightenment. Accordingly, from this approach to Schoolman's work and visual studies of the most diverse media in the face of widespread polysemy, the focus is first on understanding an aesthetic form of reason and then on its relationship with politics. The fact is that in this relationship we find reason in contemporary reality in the objective definitions of the organization of visual elements, by which the meaning of reason is associated with sociocultural, political, and economic aspects and technical and scientific progress. However, the experience and significance attributed to the different means of language and communication are diverse, especially in terms of the visual aspect. Thus, communication and the arts from the European Enlightenment through modern times to more recent history are considered for this analysis of urban contemporary art to be the political and sociocultural aspects of the past, concerning the current context of a culture with democratic values. In this sense, Schoolman's work stands out, in which the interest lies in the "reconciliation image" as the pedagogical vehicle of aesthetic education in search of democratic enlightenment. Through rereadings of Walt Whitman, Friedrich von Schiller, and Theodor Adorno, Schoolman relates the historical-social context to the period in which these works were inserted. By interpreting Whitman's respective poetry, Schoolman introduces us to the visual aesthetic quality by which he leads us to a "reconciliation image" through the work of this American poet, which refers to a visual configuration in motion. This visualization, through Whitman's poems, is significant regardless of the linguistic context. The visual images take on political forms—inclusion, receptivity, and imitation of differences—which Schoolman considers to be inclusive, receptive, and imitative images. When the object of analysis is the visual image, understanding our society's relationships requires a study that progresses far beyond the cultural aspects of the relationship between art and society. It is precisely in Schoolman's interpretation of Adorno's work that we see the need for an emphasis on critical and aesthetic theory that includes modern society in its analysis, along with the theoretical results, which achieved ample significance when ideological issues concerning capitalist society during the 20th century were discussed. However, aesthetic autonomy in Adorno's theory is characterized by the realization of a hermetic work in music, literature, and the arts.

Nonetheless, the autonomy of Adorno's theory of aesthetics was aimed at the formation and education through the hermeticism that his work offered and which he defended in his approach. Adorno's work would attract the attention of a growing audience through education, but only if his work were understood, forming a critical consciousness. However, in the general sense, Adorno's work would follow a direction in which art would cease to be art. According to him (1970), art would remain a "promise of happiness," as he reflected that it would not matter in the society of his time and that its existence would even be threatened. When dealing with rationality inserted into the universe of the arts, the critical factor was that this reason was being used to realize a work of art. This art, for Adorno, was considered a modern "work of art" precisely because it is different from the other products of the culture industry. Thus, a "modern work of art" for Adorno represented the truth content that would condition society to resist the culture industry's domination. The meaning would be in the negation of social reality, which is why his theory is known as the aesthetics of negation. However, it is in Adorno's *Aesthetic Theory* (1970) that Schoolman (2020) found the necessary evidence for the "reconciliation image," which Adorno had confined at the time to modern art, and that, in the present day, can be found in cinematography—the "reconciliation image" in film. The focus would be on the forms by which art would be produced, considering its mimetic elements—between rationality and *mimesis*—in a dialectical process for a visual and paradoxical dynamic, providing remissions of the moving image, enabling the "reconciliation image." Visual arts and cinematography can be regarded as "aesthetically analogous." Consequently, discovering the "reconciliation image" through cinema, Schoolman positively developed the concept of aesthetic reason on which he established his theory of democratic enlightenment, analyzing, for example, Rodin's *Thought* and Monet's *Rouen* "by differentiating aesthetic reason's interrelated moments of aesthetic reflexivity" and "how the process of aesthetic reflexivity works through serial images."

From the artworks and films analyzed by Schoolman (2020), it is possible to approach the democratic ideal and its relationship with the world. Accordingly, to better understand the theoretical relationship of Schoolman's "reconciliation image" related to the "imitative image," the aesthetic experience of an image in movement and of its constant transformation is the better example for the relationships between individuals and their sociocultural contexts. Simultaneously, it is crucial to contemplate the character of the manifestation of the visual artifact as a constant search for innovation through all kinds of image-based production techniques. Furthermore, it is imperative to understand the progression of images in configurations that idealize contemporary aesthetics. These developments are fundamentally processed through ideas and illusions as universal themes of creativity and are based on the achievement of greater social significance for the awareness of reality. In addition, the meaning would be determined as a reciprocation of the perceptive relations between creative individuals, artists, architects, designers, and filmmakers. For example, Jean Nouvel has asserted that his architecture was significantly influenced by film, particularly the work of filmmaker Wim Wenders. An architect is an image manipulator. For Nouvel, architecture is, above all, a production of images and of awareness through sight. Cinematographic culture teaches us to see these images in relation to time. Life relates to time, whether in displacement or in relation to buildings that, according to some rites, survive the test of time. Classic architecture is composed in a solid form, and time is presented by means of an ordered sequence (Baudrillard and Nouvel 2000). Nouvel presents the work of Wenders, which depicts scenes that expose the rhythm of daily life in a city. The importance that this filmmaker places on urban life in his cinema on technological convergence in relation to architecture has significantly influenced Nouvel's projects. The connections between the techniques used to compose the images representing public spaces have produced architecture that seeks to provide an image of both communication and appearance. In the overall composition, such a structural design offers an aesthetic of daily life. Such an image represents a civilization engaged in the struggle for rights and democracy. Many images demonstrate identity, difference, receptivity, inclusion, and, finally, some images are imitative. According to each culture, they function as visual forms of the world's diversity and, in a certain way, as a model for the continuity of the human condition. Finally, the visual narrative of the moving image is strictly related to the existential condition and the various conceptions of the world and its transformations, which, even if they do not have a clear political intention, are contained in these political transformations.

Therefore, in this analysis on "Urban Space and Collective Consciousness" of the image representing the urban context (i.e., public space), the relationship between institutional and social interdependence was discussed. Art, in this sense, was allowed without objecting or resisting forms of manifestation associated with ideals of democracy. Hence, the meaning of the current image was considered in a context of cultural diversity, situating each stage of development of the analysis. As a result, art is verified in contemporaneity through the correlations of cultural structures and urban social relations. In addition, arguments were presented for understanding freedom of expression in personal and collective experiences related to the arts in contemporary society. However, as a continuation of this research based on arts and the complex worldview, new approaches to artistic manifestation in urban space are essential, delimiting urban experience to connect abstractions with personal and collective situations of different cultural backgrounds. To this end, the following theme, "The Arts and Traces of the Ideal City," is aimed at complementary reasoning of the relationship between urban space and the autonomy of art by discussing the factors that have favored artistic emancipation in the face of science, religion, morality, and political institution.

Note

1 At that time, the art critic Alan Riding highlighted for the *New York Times* the following in a review:

> For all the natural instinct of artists to want to change the world, it is rare for their works to be displayed in a setting where government officials must take note of them. But an unusual exhibition at the United Nations complex in Geneva does just that: it forces art and politics to coexist.
>
> "Politics, This Is Art. Art, This Is Politics," *The New York Times*, August 10, 1995. https://www.nytimes.com/1995/08/10/arts/politics-this-is-art-art-this-is-politics.html

Bibliography

Adorno, Theodor. *Ästhetische Theorie* [Aesthetic Theory]. Frankfurt am Main: Suhrkamp, 1970.

Althusser, Louis. "Idéologie et appareils idéologiques d'État (Notes pour une recherche)" [Ideology and Ideological State Apparatus (Notes for a search)]. *La Pensée*, no. 151 (1970). Reprint, in *Positions*, 1964–1975. Paris: Les Éditions Sociales, 1976.

Althusser, Louis. *Sur la Reproduction* [On Reproduction]. Paris: PUF, 1995.

Baudrillard, Jean, and Jean Nouvel. *The Singular Objects of Architecture*. Trans. by Robert Bononno. Minneapolis: University of Minnesota Press, 2022. Originally published as *Les objets singuliers. Architecture* (Paris: Calmann-Lévy, 2000).

Benjamin, Walter. "Das Kunstwerk im Zeitalter seiner technischen Reproduzierbarkeit" [The Work of Art in the Age of Mechanical Reproduction]. 1939. Reprint, *Medienästhetische Schriften*. Frankfurt am Main: Suhrkamp, 2002.

Bobbio, Norberto. *Le futur de la démocratie* [The Future of Democracy]. Paris: Seuil, 2007.

Curley, Edwin, ed. and trans. *The Collected Works of Spinoza*. Princeton: Princeton University Press, 1985.

Descartes, René. *The Principles of Philosophy*. 1640. Trans. by E. Haldane and G. Ross. Cambridge: Cambridge University Press, 1911.

Garrett, Don. "Representation and Consciousness in Spinoza's Naturalistic Theory of the Mind and Imagination." In *Interpreting Spinoza: Critical Essays*, edited by Charles Huenemann, 4–25. Cambridge: Cambridge University Press, 2008.

Gulick, Robert Van. "Consciousness." *The Stanford Encyclopedia of Philosophy* (Summer 2017). Edward N. Zalta, ed. https://plato.stanford.edu/archives/sum2017/entries/consciousness/

Habermas, Jürgen. *Theory of Communicative Action*. Trans. by Thomas McCarthy. Boston, MA: Beacon Press, 1984. Originally published as *Theorie des kommunikativen Handelns* (Frankfurt am Main: Suhrkamp, 1981).

Habermas, Jürgen. The Structural Transformation of the Public Sphere: An Inquiry into a Category of Bourgeois Society. Trans. by Thomas Burger. Cambridge, MA: MIT Press, 1991. Originally published as Strukturwandel der Öffentlichkeit. Untersuchungen zu einer Kategorie der bürgerlichen Gesellschaft (München: Luchterhand 1962. Reprint, Frankfurt am Main: Suhrkamp, 1990).

Hegel, Georg Wilhelm Friedrich. *The Phenomenology of Spirit*. Trans. by Terry Pinkard (Cambridge Hegel Translations). Cambridge: University Press, 2018. Originally published as *Die Phänomenologie des Geistes* (Bamberg and Würzburg: Joseph Anton Goebhardt, 1807).

Hegel, Georg Wilhelm Friedrich. *Lectures on the Philosophy of Art*. The Hotho Transcript of the 1823 Berlin Lectures. Trans. by R.F. Brown. Oxford: Clarendon Press, 2014. Originally published as *Vorlesung über die Philosophie der Kunst*. Berlin 1823. Rewritten by Heinrich Gustav Hotho (Hamburg: Felix Meiner, 1998).

Horkheimer, Max, and Theodor Adorno. *Dialektik der Aufklärung* [Dialectic of Enlightenment]. Amsterdam: Querido, 1947. Reprint, Frankfurt am Main: Fischer, 1998.

Houlgate, Stephen. "Hegel's Aesthetics." *The Stanford Encyclopedia of Philosophy* (Spring 2016). Edward N. Zalta, ed. https://plato.stanford.edu/archives/spr2016/entries/hegel-aesthetics/

Kant, Immanuel. *Kritik der praktischen Vernunft* [Critique of Practical Reason]. Riga: Johann Friedrich Hartknoch, 1788. Reprint, Stuttgart: Reclam, 2008.

LeBuffe, Michael. "Spinoza's Psychological Theory." *The Stanford Encyclopedia of Philosophy* (Spring 2015). Edward N. Zalta, ed. https://plato.stanford.edu/archives/spr2015/entries/spinoza-psychological/

Lyotard, Jean-François. *La condition postmoderne: rapport sur le savoir* [The Postmodern Condition: Report on Knowledge]. Paris: Minuit, 1979.

Marès, Mathilde. "La liberté est-elle une illusion ?" [Is Freedom an Illusion?].
Philosophie Magazine, March 3, 2013.

Nadler, Steven. "Baruch Spinoza." *The Stanford Encyclopedia of Philosophy* (Summer 2020). Edward N. Zalta, ed. https://plato.stanford.edu/archives/sum2020/entries/spinoza/

Orwell, George. *Nineteen Eighty-Four. A Novel*. London: Secker & Warburg, 1949.

Riding, Alan. "Politics, This Is Art. Art, This Is Politics." *New York Times*, August 10, 1995. https://www.nytimes.com/1995/08/10/arts/politics-this-is-art-art-this-is-politics.html

Schoolman, Morton. *A Democratic Enlightenment: The Reconciliation Image, Aesthetic Education, Possible Politics*. New York: Duke University Press, 2020.

United Nations Gifts. "Mankind's Struggle for a Lasting Peace." March 19, 1953. Accessed September 20, 2022. https://www.un.org/ungifts/mankind's-struggle-lasting-peace

Weber, Max. *Wirtschaft und Gesellschaft: Grundriss der verstehenden Soziologie* [Economy and Society: Outline of Understanding Sociology]. Berlin: Peter Kietzmann, 1922. https://www.textlog.de/weber_wirtschaft.html

3

THE ARTS AND TRACES OF THE IDEAL CITY

The image in the urban context was discussed concerning the interdependence of social structures in the previous chapter, "Urban Space and Collective Consciousness." The discussion highlighted the image's meaning in cultural diversity from modern to contemporary times, delimiting the sense of freedom of expression. Now, continuing the analysis on arts in public space in "The Arts and Traces of the Ideal City," the focus moves to the relationship between public space and the autonomy of art. This chapter discusses the factors that favored artistic emancipation, its relationship with social reality, and democratic values. Therefore, the theory of beauty[1] is the starting point for situating oneself in the system of ideas, referring to life in society, the world of arts, by the aesthetic experience. The notion of the beauty ideal is, in principle, a way to imagine perfection above any mundane reference, associated with divinity and the unachievable ideas of truth and justice (Jäger 1947; Gurley 1999). Moreover, beauty is the most significant motive for artists to produce a beautiful masterpiece and also to revert that sense by giving up those respective achievements, reduced to a disillusioned search for the ideal of beauty. Thus, artistic representations in correspondence with everyday life, linked to their imaginary and real experiences, start to consider art as a representative means of reality through the satisfaction of wills, immediate pleasure, and sometimes reversion to beauty. These representations can even highlight the horror. How does one interpret artistic representations' diversity in the face of the many references found in the rhythm of social development, proper to cities' dynamics in their evolution? To think of art, culture, and reality, emphasizing aesthetic theories, implies a rereading and reinterpretation of Western and non-Western knowledge history (Auerbach 1994; Rampnoux 2008). However, when focusing on the sense of beauty, art, and reality, it is fundamental to remember that religion or myth, philosophy, and science

DOI: 10.4324/9781003314943-4

represent the three forms of thinking that seek sense for the world. Since their Western origins, these forms of thought, from the mythical or symbolic form to rationality, have been a means of understanding human beings and defining the ways of thinking, each with its history and meaning, since ancient Greece. To this aim, the leading question in this chapter is as follows: Would democracy occur through the actualized context of programmed or spontaneous sociopolitical images or through different artistic ways of creating the same content that has always had the same effects on culture?

Through an aesthetic-philosophical reflection, the first part of this analysis, "The Arts and Their Significance to Urban Space," focuses on a problem specific to the discussion with the essential aspects of art and its effects on the development of public space. Therefore, the foundations of Western knowledge related to the arts, urban space organization, and interconnected socio-educational relationships are discussed. Next, the section "Lack of Beauty and its Conflicts" emphasizes the artistic achievements observed in analogies through an aesthetic reflection on the relationship between beauty and horror displayed in public space. The arts are presented in a relationship of interdependence from the first manifestations of the Ancient World, throughout Greek antiquity, through the Renaissance and modernity to the present day, based on formal coincidences. Artworks are compared in their predominant characteristics in the face of different historical contexts to think about the nature of these expressions, and also about how human beings reproduce themselves in behaviors and feelings in a mimetic process through artistic manifestation in their evolution. In "Beyond Illusion: A Reality for the Collective Experience," it is historically evidenced that the first narratives dissuaded truth and reality as a strategy, in this case a method of warfare with reports to mislead or distract attention, which alludes to what is now understood as fake news. Finally, focusing on the origins of urban centers, the aim is to discuss the values operated through crises and new governmental achievement (Morin 2007), namely instability and insecurity when seeking security, establishing optimistic and innovative forecasts by presenting illusory images and narratives.

The Arts and Their Significance to Urban Space

The first reflections on life in society are found in Plato's *Republic*, where in Plato's aesthetics, "poetic"[2] activity is associated with the sensitive and what belongs to it: Sensuality, passions, emotions, and affections among others—meaning everything that concerns artistic activity related to society. Now, as far as philosophy is concerned, it is considered the intelligible, the knowledge, and the spirit. Between art and philosophy, there is necessarily a relationship of subjection, distancing reason and *logos*[3] from the sensitive universe. For Plato, art would be in this condition, submissive to the organization of the State. Plato describes artistic activities in all of their perversion capacities and effects on the soul: The spirit—one's conscience—which implies the moral notion of what is right or wrong, influencing human behavior itself in social life. Consequently, in theory, art would benefit the spirit or power of evil. Plato perceived the consequences that evil through artistic achievements could bring to the city, Athens society, the classical period, and the 5th and 4th centuries. According to Plato, this evil, mediated by the arts, could be avoided if they were not conducted according to the principles judged by him to be of political order and reform, through education or artistic education, to guarantee a fair and harmonious city. Plato was influenced by Socrates' ideas of life's essence, the soul, which would belong to the universe of truths, good, and beauty—a conquest by wisdom and knowledge. However, when the soul is embodied in daily life, it would always be searching for the pure, eternal, and immortal condition compared with the physical and worldly form. Nevertheless, this search becomes a continuous task amid worldly accomplishments, and one can understand that, according to these philosophers, influences could approach this encounter with wisdom and knowledge, especially when it comes to the arts aimed at the general public. At that time, artistic realization was related to tragedy and comedy, spoken word, poetry, and music. In this way, Plato intended to restore Athens in the face of decadent governments through a rigorous system of education. For example, see the context in which Plato created the *Republic*. In 431 b.c.e., the Peloponnese war began; Plato was between 2 and 4 years old, and the war lasted almost 30 years. Thus, the author of the *Republic* grew up during the conflict with Sparta in Athens, which ultimately lost its empire, supremacy, and power. After its defeat, Athens never again regained the glory that it had before the Peloponnese war. Sparta imposed an oligarchic regime on Athens and ended Athenian democracy in 404 b.c.e. using tyranny, eliminating the opposition, which included politicians, aristocrats, and citizens. Among the intellectuals who were deprived of teaching was Socrates. As reported by Plato, Socrates was condemned to death in 399 b.c.e., but he perceived it as a form of "freedom" from the physical world for the evolution of the soul and as a teaching to his followers. One painting that illustrates this historical account is *The Death of Socrates* by Jacques Louis David (1787). In his work, Plato, in the form of dialogues, describes the ideas of Socrates, who left no written texts. Plato is not a part of these dialogues. The posthumous interpretations of his work seek to distinguish Socrates' ideas from those of Plato himself. Plato's work was systematized at the beginning of the Christian era. Thus, in the Middle Ages, Christianity identified itself with Platonic thought and the idea of an essence on a plane above worldly reality.

Timelessly, we could say that Plato develops an idealistic and receptive aesthetic of the effect, reflecting on art concerning society and even the psychology of art. Plato lived in a time of intense political and cultural crisis. His project for a good city was to solve Athenian decadence and democracy. Throughout his life of approximately 80 years, Plato's philosophy, characterized by a diversified and even contradictory thought, finds in the theme art and society the references of Western thought. His philosophy was either through the influence of ideas, passages that deal with beauty, or even society's encounter and art's effect, regarding the city's problems: Education, politics, and the *paideia*—art pedagogy, culture, and the city's constitution or government. The discussion about education is part of the *Republic*, Book 6, where the well-known *Allegory of the Cave* is included. The allegory concerns Plato's articulation of the soul's path to philosophical enlightenment, away from the pacifying darkness of a cave, dominated by appearances, in analogy to shadows and rhetoric, to the encounter of sunlight, the metaphors meeting freedom and truth (Jäger 1947; Gurley 1999).

Plato was concerned with art education to enable all citizens to live in harmony in a state free of tyranny. Hence, parallel to Plato's aesthetics, part of *Republic*, he develops a pedagogy and politics through artworks that can offer beauty, a concept by which he formulates essential questions for defining beauty and artistic practice. However, for its project of a harmonious state, intransigence with the arts is mainly due to the arts of appearances, constituted by imitation. Imitation is the sense of the art of mime, of simulation, producing false impressions, the simulacrums that divert attention from both concrete reality and the essential, which are a single reality. Thus, Plato deepens the differences between the arts by prioritizing the forms of *mimesis* in artistic realizations. Plato respects beauty, but he distrusts art for moral and political reasons and for the importance of art, which he recognizes: Art is far from being an insignificant subject. Besides the universe of ideal beauty, which aspires to meet transcendence and the divine, there is the world of the concrete experience of art, lived by the individual and society—on the one hand is absolute perfection, the sphere of beauty, and on the other hand is the sensitive world, imperfect but perceived by aesthetics, in philosophy. The reflection on art for Western knowledge is presented up to Romanticism in the 19th century with long contradictions of Platonic theories and interpretations of the theory of beauty and *mimesis* renewing it and finding as a point of reference, throughout the philosophical tradition, the theories of Aristotle (384–322 B.C.E.), who contradicts Plato's theory of beauty. Aristotle favored imitation, *mimesis*, and refused to separate the intelligible world from the sensitive world, instead associating pleasure with the artistic imitation of nature. He was against the submission of art to the authority of philosophy and politics and did not favor the exclusion of any form from the city that was unsuitable. For Aristotle, art is beneficial to both the individual and society. In all arts, *mimesis* concerns the imitation of nature and the representation of reality. Tragedy as the main conception of art prevailed for many centuries and was called "mimesis." From antiquity to the Middle Ages, the conception of reality and its representation of the ancient and medieval worlds were distinguished, according to Erich Auerbach, between the Odyssey and the Bible. While the real and the everyday are not absent in Homer's work, the Bible instead offers spectacle and figurative interpretation. It is the life of Christ related to daily life, the family, and the sublime tragedy, marking a categorical rupture with antiquity in the sense of *mimesis* in his literary conception of the real (Auerbach 1994).

Since the Renaissance, the translation and interpretation of the term "mimesis" have founded important philological and theoretical debates and played a crucial role in the history of artistic thought. In short, the theory of *mimesis*, itself associated with the term, has its meaning etymologically in ancient Greece, with distinct purposes of interpretation in Latin translation. This was added to the values of the Middle Ages, which is when the passage from the term *mimesis* to *imitatio* and *imago* occurred, designating the image of God, associated with all segments of the sacred. The expression *imitatio Christi* from the 14th century illustrates this transformation well. The idea of man being created in God's image places the term *imago* not only in the sense of a copy but also that of an analogy. From the Latin meaning, the Italians developed the theory of imitation. However, in opposition to this theory, the French adopted the approach of Aristotle, namely *mimesis*, which has prevailed in European thought since the 17th century. From this French conception, the Germans developed their reflections on *mimesis* (*Nachahmung, nachmachen, kopieren*, and *nachbilden*) in the 18th century. In the 15th and 16th centuries, during the Renaissance, Aristotle's theory of *mimesis* was referenced in painting as an imitation of nature. Art, in its origin, has the sense of technique, given to the term *téchne* in ancient Greece. The term reached, at that time, a much broader spectrum than today. The word *téchne* formerly had a meaning both for manual work and for art or military strategies. Thus, it manifested itself as a technique not only for manual work (of slaves and craftsmen) but also for mental and intellectual work (of free men), such as the liberal arts (*quadrivium* and *trivium*). From the 11th century on, the Latin word *ars* meant skills or activities associated with the technique, like the Greek term *téchne*.

During the Renaissance, the dominant aesthetic principle was imitation (*mimesis*) by mathematics and geometry in the field of art, not by copying but through the techniques of adaptation as submission to the techniques of perspective. The objects of art were nature, human beings, and God. Yet, in designating "the human being," it was not yet about the human being or subject; therefore, there was no mention of subjectivity at the time. At the beginning of the Renaissance, creative capacity did not confront the divine conception. This power of creation as the notion that knowledge does not arise from anything, but rather from the knowledge acquired through science, emerged with Leon Battista Alberti (1404–1472), a painter, sculptor, musician, and architect who developed perspective as the norm in the Renaissance according to the painters of Quattrocento. Both pre-Socratic knowledge, whether of the Pythagorean school or of Thales, and references to the works of Plato and Aristotle allowed humanism to triumph with reason and sensitivity. Humanism became the measure of the creative act of an artist, the interpreter of nature, according to Leonardo Da Vinci's formula. *The Vitruvian Man* (*L'Uomo Vitruviano*) is a work from 1490 based on a work of architecture older than the famous Vitruvius. It mentions the perfect divine proportions; therefore, this man would be the human ideal. All of the work has dimensions based on the number *phi* (1.618) that the Greeks divulged.

The School of Athens (1509–1511) is not only a painting masterpiece by Raffaelo Sanzio da Urbino (or Raphael) but also the fundamental meaning of classical thought. It represents the influence of The Academy, Plato's School of Philosophy, which existed around 387 b.c.e. in Athens. The region where it was founded, during the archaic period, was a public park called Academy, which was devoted to the hero Academos. In this place, Plato founded his school of philosophy on a small property near Ippios Kolonos and developed his philosophy. Even after his death, his followers continued, and Plato's school functioned for approximately three centuries during the late classical and Hellenistic periods. Among its best-known students were Aristotle, Xenocrates, Heraclides Ponticus, Eudoxus of Cnidus, Polemon, Arcesilaus, Carneades, Philo of Larissa, and Charmadas.

Since the Renaissance, with a humanist ideal, Raphael's *The School of Athens* has represented human beings' position in the full domain of their reason, searching for more excellent knowledge for their beliefs. This masterpiece demonstrates the composition's evident rigor guiding the leading figures in the center of the painting, representing Plato and Aristotle discussing their main ideas—opposites that drove all Western thought. The gestures represent their opposition to Plato's ideas, who points with the finger of his right hand to an order of ideas (Plato points up) up to the *aletheia*, meaning "no mistakes" in archaic Greek. A path that starts from the science of numbers to the domain of music, and with this domain a cosmic harmony and divine beauty, is to be elevated to the highest categories: The truth, the good, and beauty. In his left hand, Plato holds the book of his work, *Timaeus*. Aristotle, however, has an earthly and physical vision (his right hand points down) based on ethics in the work *Nicomachean Ethics* (a book he holds with his left hand). Plato and Aristotle are the two leading representatives of classical thought. This masterpiece of Raphael is essential in depicting not only the moment of the affirmation of humankind in their world but also the fundamental, which is rational truth, through knowledge. This painting celebrates the classical formation, the seven liberal arts—*Artes Liberales*—in the *quadrivium* (four liberal arts): The mastery of mathematics, the science of numbers, understanding arithmetic, geometry, music, and astronomy—and the *trivium* (three liberal arts), the dialogue domain, which comprised grammar, rhetoric, and dialectics. Art has its autonomy when it ceases to be a mechanical, manual, artisan art and becomes recognized as an intellectual activity, which has been a long emancipation process since the Renaissance. Progress was made through two critical phases, the first of which was in the 17th century through the beauty of detaching from the values of good and truth, in the Platonic sense. Then, at the end of the 18th century, the imitation of nature, "of beautiful nature, in the fine arts," ceased to be the artist's only purpose. These transformations did not arise by themselves; they had been consequences of social, economic, and political conditions since the Middle Ages. It was these transformations that allowed the new conceptions to materialize as a reality. For example, artists' social recognition, distancing themselves from the artist's status, must be associated with artists' progressive distancing from religious, monarchical, and aristocratic tutelage.

The autonomy of art appeared as the result of many factors that favored the emancipation of art from science, religion, morality, and the political institution. The Renaissance represented only an essential stage for the whole process of art's autonomy, which continued through the Enlightenment, the absolutism of

Louis XIV, and the 18th century with its aesthetic autonomy, to find a favorable context for emancipation and establish the status of the artist. With all of the arts' transformations in the *Cinquecento*, later during the 16th century, the first art academies in Europe emerged, and the canons of the fine arts academies were established. The main references were the Royal Art Academies of France and England. It was when the aesthetic discipline appeared in Germany with Alexander Gottlieb Baumgarten (1735), in the 18th century, that art became the object of study of not only philosophers but also artists and the public. They attended the first painting and sculpture halls. They knew the canons of the artistic universe through the notions, concepts, and categories offered and according to which artistic achievements were oriented. On the one hand, this aesthetic autonomy favors the autonomy of the subject by his or her faculty of aesthetic judgment, freely expressing his or her taste for natural or artistic beauty. However, one can understand that an aesthetic discipline's autonomy means the delimitation of a space for reflection on art, as access to the idea, truth, meaning, and absolute, as much as science and philosophy are concerned, which little by little has freed itself from the old rules and dogmas. The specific space for reflection and artistic criticism—aesthetics—can be a risk if it is distant from everyday life (Jimenez 2005). While it finds the freedom to establish its own rules with an autonomous sphere, it would be distant from the reality of everyday life—moral, ethical, political, and so on—to the sense of art as a reality. Thus, aesthetics isolated from reality in its autonomy would condition the reflection on art to lack principles, making it unimportant as knowledge. The arts' value as an expression of human liberation in the history of art had significant importance in modernity. All of these events in the arts were part of a political and social transformation period, favoring the new ideas and behaviors of artists and thinkers. At the end of the 18th century, the bourgeoisie progressively conquered its self-assertion, mainly with the advent of the Glorious and French Revolutions and the progressive conquest by the public space with commercial and industrial development, factors that were favorable to the start of the Industrial Revolution. Thus, modern art emerged with the desire to break away, not from content, but from traditional, conventional, and academic forms. Nevertheless, the understanding of human reality, of Western civilization in the past 2,500 years, is guided by the natural and rational phenomenon—as philosophy and science have emerged—and no longer by myths (Rampnoux 2008). Since then, through great thinkers' ideas—first philosophers and then scientists—humankind has reached the 21st century. During this process of autonomy, globalization, the ideal of democracy, and humans have found themselves in the universe of art and culture.

Lack of Beauty and Its Conflicts

Despite all of the abstraction for understanding beauty, considering it an essential part of our faculties, it is in daily life that individuals conduct their activities. It is also in daily life where they build their contemporaneity and the diversity of paths, overcoming and resigning themselves to the challenges with the know-how for the present's questions, reviewing the past and always imagining the future. Yet, we can state that it would no longer be possible to perceive a change over time. Memory itself offers us the construction of history by establishing places through which we can review moments in time or even search for paths, often making new choices. In reality, the individual inhabits a cosmopolitan life with a cultural diversity between choices, leaving only the values of morality, ethics, and public life—in short, the rules of social life. In all cases of social life, the rules lead us to the following questions: By whom, for what, and where are they established? If we already find the answers in the works consecrated by science, for example, that the system knows only its survival condition, we would have the resignation and acceptance of always being dependent on any fact. Thus, would the individual have the freedom to innovate? We understand freedom as free will and choices; we are the masters and drivers of the system through cultural influence, which promotes innovation and desire concerning the ideal of freedom. Such an ideal is the beauty felt toward "truth," the pursuit of perfection, or the imbalance with the imagination. Then, reason leads us to the sublime, the constant artifices and inventions that always remind us of our fragility in the face of the greatness of nature.

The technical or artistic values that idealize beauty and the sublime opened a new direction for the fine arts. Since then, different arts, each with their own specificities, have been understood as a whole, creating art regarding nature—its representation. Art, in this sense, became a universal concept. Art evolves, each specificity in its own course, searching to create new art through technological development or the technical domain, depending on creativity. Even if novelties, inventions, and artifices can always arise, theories also find a new reason for and systematization of achievements. The arts' views discussed have developed according to artistic creation and vice versa, maintaining a relationship of interdependence. Yet, from the first manifestations of the ancient world, or throughout Greek antiquity, even in the Middle Ages, the Renaissance, modernity, to the present day, how many coincidences can we evidence for the creative act? See, for example, the list of expressions when we compare artworks from different historical periods in their predominant characteristics. Thinking about the nature of these expressions, about how human beings reproduce in mimetic process behaviors and feelings in their evolution, has its sources in their origin through artistic manifestation. However, from scientific and methodological points of view, we already have some registers in Darwin's (1872) studies:

> The ancient Greek sculptors were familiar with the expression, as shown in the statues of the Laocoon and Arretino; but, as Duchenne remarks, they carried the transverse furrows across the whole breadth of the forehead, and thus committed a great anatomical mistake: this is likewise the case in some modern statues. It is, however, more probable that these wonderfully accurate observers intentionally sacrificed truth for the sake of beauty.

Such initial studies were fundamental to the development of psychology, and consequently, to that of aesthetics, perception studies, iconography, semiotics, and the language of the arts as well. However, one can follow some analogies through an aesthetic reflection on the relationship between beauty and horror. For example, the works of Ernest Pignon-Ernest from 1988 to 1995 in Naples, *La peau des murs* along with *David et Goliath*, challenge the people in a public space that keeps its history with Greek, Roman, and Christian mythologies. However, this French artist's artworks are not limited to these mythologies; with interventions in the public space, he appropriates other images from consecrated paintings, such as *Pietà* (1498–1499) by Michelangelo Buonarroti, located in St. Peter's Basilica, Vatican City. The *Pietà* is perhaps the best-known masterpiece and has been appropriated in many other artistic creations. Representing Jesus dead in his mother's arms, it is one of the most famous Renaissance sculptures. Pignon-Ernest's urban art, the *Pietà sud-africaine* (*South African Pietà*), displays the image of a person in existential despair or, again, in situations that have marked many artistic achievements, offering *catharsis* through piety. The *Pietà sud-africaine* is an artistic work to make viewers think about Durban, South Africa. Pignon-Ernest not only used the scene that sparked the Soweto uprising during the Apartheid era, on June 16, 1976, with an image of the death of a 13-year-old boy named Hector Pieterson, in the arms of a man—but he also transformed the young man into a dignified person carrying a boy dying after being shot by police. Pignon-Ernest distributed images of *Pietà sud-africaine* in Durban, because it is a port area with a massive flow of people. The function of his art, in this context, was to make humanity aware that not only was Apartheid a brutal reality but also, in the same way, AIDS was a reality spreading in the same proportion, mainly in this region. Even earlier, in this same line, the artistic work *Dead Child* (1944) by Candido Portinari stands out. Portinari showed the influence of Picasso from the 1940s, after the vision of *Guernica*, offering more drama, expressing tragedy and human suffering. Thus, it acquires the character of denunciation concerning Brazilian social issues, revealed in works such as *Northeastern Migrants* (1944) and *Dead Child* (1944), thus becoming a recurring theme in its production expressionist elements. The artist expresses the northeastern migrants' tragedy through the crisp gestures of their hands and tears of stone.

Another work that explores the symbolic aspects of the figure who fights for a social or political ideal, defending the ultimate cause and even renouncing or dying for it, is the martyr character. In this sense, we associate the Christian repertoire with the crucifixion, dying for faith instead of renouncing it. Jean Genet, represented in Pignon-Ernest's artwork, does not directly quote the novels but presents a tension between sacred and profane through an aesthetic shock. In addition, it takes the expression of the scream, concerning the appropriation of Edvard Munch's painting *The Scream* (1893) or *Laocoon* (40–20 b.c.e.) by Pignon-Ernest. His artistic works have been elaborated throughout his artistic career for the benefit of theater, cinema, and social struggles. Pignon-Ernest's posters allow us to retrace the encounters and struggles that marked his work. He frequented the Fluxus circle in Nice and André Benedetto's militant theater in Avignon in the 1960s. His posters testify very early to his reflections on the need for an art that is in direct contact with reality, denouncing the injustice of his time. *Urgent crier* (1966) became one of the mottos of his work, both through his urban routes making works of art of places and realizing more punctual posters in which he associates texts and striking images.

To find a sense of beauty in the image's configuration is highly complex, and the meaning of art regarding cultural context is diverse. In this complex relationship of expressive capacities, of relations of various formats, it is always useless for us to limit ourselves in front of the immensity that the universe of actualities has become. We can only console ourselves knowing that images configure the new and that surround us more each time. In general, in its creative processes, art is always looking for a criterion for evaluating possible creativity, linked to the diversity of realization and originality. As far as completion is concerned, it provides concrete proof in analysis and evaluation for its material presentation. Mastery of originality is necessary for evaluating the accomplished artwork and skill of all existing elements and the diversity of the present's numerous other achievements in a retrospective sense.

Beyond Illusion: A Reality for the Collective Experience

As we have known for a long time, our illusion of reality through the arts is due to the influence exerted not only by the current media technological resources of mass culture but also, even early, by the ancient forms of reproduction, and first by oral language. We can discuss not merely an influence originating in the ideal of domination of public space by the bourgeoisie but rather what represents, in the general context, a narrative as influence through literature and art. For example, it would already be used as a propaganda method in the year 1 b.c.e. in public space. The emperor Augustus commissioned Virgil to create an epic that would narrate Rome's rise and, above all, build a sacred image of Augustus as the great emperor of Rome, crowning "universal peace and happiness." With this epic, the illusion through the narrative of the Latin poem the Aeneid, we can perceive how feelings being represented in a particular way can touch the people and involve them, as it has since the first great civilizations. See, for example, even the figure of Laocoon, which in ancient Greek means "he who understands the people." Laocoon represents an essential character in Virgil's Aeneid, Book 2. In this epic, in Laocoon's story, Poseidon's priest––Laocoon––warned the Trojans about a wooden horse found one morning on the beach, which they thought was an offer from Poseidon. The Trojans, some in doubt about the horse's origin, wanted to burn it, while others believed it was a gift and took it into the city. Virgil presented Laocoon as one who understands the people, warning them by saying the following in Latin: "*Equo ne credite, Teucri. Quicquid id est, timeo Danaos et dona ferentis*," which was translated into English by Theodore C. Williams (1910) as follows: "Trust not this horse, O Troy, whate'er it bode! I fear the

Greeks, though gift on gift they bear" (Greenough 1900). In Virgil's narrative, at the moment Laocoon suspects that it could be an offering from Poseidon, a colossal snake emerges from the high seas, attacking his two sons and then Laocoon himself. This image of horror witnessed by the Trojans leads them to believe that Poseidon's revenge is because of Laocoon's contempt for the present. However, it is here that we encounter historical evidence that the first narratives employed a strategy that ignored truth and reality. In this case, the war tactics involved narratives that aimed to divert the readers' attention or deceive them, or what we would now understand as fake news. It is a scene of great injustice and tragedy for a city and a people who deluded themselves with a gift, a people who did not believe in the one who would understand it, the Laocoon—*laos* (people) and *koêô* (understand). Not believing in images means, first, believing in another reality, initially dissociating oneself from tradition and the conventional, eliminating any nostalgic feeling—but then, naturally, there is a need for individuals to believe in a new reality. Second, it means believing in understanding all forms of dissimulation of society. In our day-to-day practice, we experience images from all over the world. They lead the individual's consciousness to social reality as the reason why cultural production finds meaning in the natural process of the demand for the consumption of images. Capitalist society has, by definition, the superficial, by consumption, by the value of use, referring to the durability of products, their usefulness, waste, and the exploitation of reserves given their survival. When we think about ideologies, the notion of social equality is seen today as the pursuit of pleasure by individuals. Being satisfied is, above all, the *sine qua non* of citizens in consumer society. In this sense, the idea of equality is understood as the desire to acquire the right to satisfaction. Thus, the democratic principle in contemporary society, or consumer society, is to achieve equality in the face of the image of social "success." This image is no longer merely the right of expression. It is, of course, the force of capitalism's ideology. In this sense, knowledge through experience keeps us in history.

In the 21st century, humanity presents itself in its cultural diversity as a planetary reality under the tutelage of technology and science. Yet, despite such facilities, closeness, and the single reality, humanity does not cease to be diverse. However, in its continuity, there is the need for an illusion for creations, which so many other artistic achievements have made possible. Nevertheless, the belief that illusion is essential divides opinions. Therefore, at this stage, we do not know in experience the importance of not having it. Let us remember that Plato conditioned the purpose of art to the *paideia*,[4] the Greek educational curriculum. In reality, art has always participated in public life, and artists have always formed a reality for the collective experience or public space, even if it is an art offering the illusion. Nevertheless, Morin (2007) stated that even if many people may believe they would lose everything if they lost their illusions, it is a mistake in the current planetary crisis. For this thinker, we would make a prodigious conquest by losing our mistakes, with the necessary conscience remaining—the real reason to decide the future. Thus, we understand that this consciousness of "being in play" (*ludere*) of mistakes and successes, that is, in its sense, by illusion (also *ludere*) as a game in its default parameters with possibilities of errors and successes, of mistakes, and finally of illusions by illusions, would be far from the reason for the truth (*Aletheia*). Morin addressed the loss that humanity experienced with the promise of progress, yet at the same time, it was significant progress to discover that progress is a myth. Furthermore, what is perceived is a system of values that operates through instability and insecurity and seeks security through establishing optimistic forecasts. One can perceive this dynamic in industry and commerce, universities, arts, and politics, exemplified throughout this analysis. However, all of that relationship with beauty, the perfection values of life, the image of success, and the search for innovation consists of perception (aesthetic), communication, and reality. The aesthetic experience of some artworks was consecrated by time, making them masterpieces. Regardless of the aesthetic criteria of evaluation or the judgment of the taste of masterpieces that made the history of Western art by experts, their eternally present values are significant. Still, cultural significance is of the most outstanding importance for undergoing social changes. These are works that cross time and continue to be contemplated, touching their observers' sensibilities, a public not specialized in art, but who perceive the values intrinsic to their human nature, by feeling, in the image. Therefore, we still require an understanding (*Verstand*) in the Kantian sense that it is possible to understand an object of art, comprehending it through the senses. The aesthetic experience in Transcendental Aesthetics has the sense of perception (*Perzeption*). If a work of art has the power to reach our feelings—if perceived, then one can think about the object and understand it according to each individual's repertoire and then conceptualize it. The concept (*Begriff*) and perception offer the individual sensations (*Empfindung*) that influence the senses (*Sinnlichkeit*), ultimately making it possible to awaken their interests or desires for the appearance (*Erscheinung*) of the work—apparition in the sense of a phenomenon. On the one hand, this analysis was concerned with what exists in the world and how to relate to human beings' paradoxes, complexities, and idiosyncrasies through art. On the other hand, there have always been questions seeking the relationship between all dichotomies, especially between good and evil, to find the best solutions on a moral and ethical basis. However, there have only been different forms of answers for the same things in various configurations. If these answers solve things globally, then it means a constant reconfiguration between errors and successes, between good and evil. Nevertheless, faced with the complex relationships of the phenomena of human existence in an innovative sense, this analysis looked for the Western tradition to think about "innovation" in the face of so many configurative elements that can be found in space. It was not a specific space for the contemplation of artwork, such as a museum or gallery, but rather the public space that this analysis focused on. This space presents all aspects of daily life. Throughout the history of artistic creations, since antiquity, even with the transformation of perception, any individual, initiated or not, could have feelings for art and be affected by arts—whether by Michelangelo's *Pietà* or the contemporary values present in the public space, or by Pignon-Ernest's urban art in *Pietà sud-africaine* or in *Dead Child* by Candido Portinari.

Art has transformed, in its history, the way that it can be understood. The technical evolution and the importance of scientific status do not matter. The metropolis's rhythm has contributed to raising art such that it enjoys an excellent reputation. From prehistory to contemporary society, the orientation of artistic speculations and the transformation of lifestyles offer us a tremendous inventory of all of the achievements that would not have space, presented here. Therefore, in general, what represents the configuration of a contemporary image, in its search for a balance between good and evil, between beauty and horror, is the innovation, improvement, and perfection of the sense of beauty in each new configuration. The dynamics of global metropolises, in their process of transformation, always seek to meet the needs of *modus vivendi* for a better condition. That is not only through the means already discussed, which alone would lead to other reflections of the constant artistic movements, systemic and market adjustments, production and work relations, and new products and services. However, that is the representation of a standard civilization: Everything, in synthesis, is the consequence of standardizations in the configuration of a single image.

Art is faced with technological evolution and science, in a scenario of contemporary universality, which is not only that of social transformations but also, above all, of the environment. Sensitivity, with greater intensity for some artists, less for others, is not in the context or at the time, but in the human species through time, and what is done, with the feeling, is that it is related to a context. One can do beautiful or horrible things. The method of doing good and beautiful things is how the creative processes and intentions, techniques, and materials were used. Art has always depended on technique in its development, which results from knowledge, learning, and the ability to transmit and express spontaneously or intentionally what is perceived in the environment. Art is the reproduction—*mimesis*—of balance, reflection, understanding, patience, happiness, joy, love, compassion, beauty, and good feelings, as well as imbalance, irrationality, pain, unhappiness, and despair—in short, horrible feelings. Good or bad, we perceive these feelings' relationships in the sensitive world, with personal experiences. Emotions and their interweaving result in artistic achievements and expressions are configured with similarities, even between artworks situated in distinct and distant contexts and times. Historical records support the notion that the origins of writing and visual language arose through an association between the symbolic representations of needs and objects within the environments in which our ancestors lived. Their drawings and lines indicated simple ways of transmitting information and ideas, approaching a written language system. Thus, the technical evolution of communication has been the main driver of civilization, the *polis*—or city in Greek (Caves 2004, 520), and the possession of knowledge. In this sense, the possibilities for the continuous development and evolution of the urban environment through the implementation of creativity, techniques, and learning processes through the arts are discussed in Chapter 4, "Democratization of Art and Culture."

Notes

1 Crispin Sartwell:

> It is a primary theme among ancient Greek, Hellenistic, and medieval philosophers, and was central to eighteenth and nineteenth-century thought, as represented in treatments by such thinkers as Shaftesbury, Hutcheson, Hume, Burke, Kant, Schiller, Hegel, Schopenhauer, Hanslick, and Santayana. By the beginning of the twentieth century, beauty was in decline as a subject of philosophical inquiry, and also as a primary goal of the arts. However, there were signs of revived interest by the early 2000s.
>
> "Beauty," The Stanford Encyclopedia of Philosophy (Winter 2017), Edward N. Zalta, ed., https://plato.stanford.edu/archives/win2017/entries/beauty/

2 Nickolas Pappas:

> In all Plato develops three theses during this first half of Book 10: Poetic mimesis, like the kind found in painting, is the imitation of appearance alone, and its products rank far below truth (596e–602c). Therefore poetic mimesis corrupts the soul, weakening the rational impulse's control over the person's other drives and desires (602c–608b). It should therefore be banned from the good city.
>
> "Republic 10: copy-making," "Plato's Aesthetics," The Stanford Encyclopedia of Philosophy (Fall 2020), Edward N. Zalta, ed., https://plato.stanford.edu/archives/fall2020/entries/plato-aesthetics/

3 From Greek philosophers, logos become a philosophical concept translated as reason, as much as the capacity of individual rationalization or as a cosmic principle of order and beauty. The logos can also be seen as the "reason" of all things, being the cause that explains the human existential longing so discussed in philosophy.

4 Jack Crittenden and Levine Peter:

> The ancient Greek city state or polis was thought to be an educational community, expressed by the Greek term paideia. The purpose of political—that is civic or city—life was the self-development of the citizens. This meant more than just education, which is how paideia is usually translated. Education for the Greeks involved a deeply formative and life-long process whose goal was for each person (read: man) to be an asset to his friends, to his family, and, most important, to the polis.
>
> "Civic Education," The Stanford Encyclopedia of Philosophy (Fall 2018), Edward N. Zalta, ed., https://plato.stanford.edu/archives/fall2018/entries/civic-education/

Bibliography

Auerbach, Erich. *Mimesis: dargestellte Wirklichkeit in der abendländischen Literatur.* 1946. Reprint, Basel: Francke, 1994. *Mimesis: The Representation of Reality in Western Literature.* Trans. by Willard R. Trask. Princeton and Oxford: Princeton University Press, 2003.

Benedetto, André, and Louis Pons. *Urgent crier* [Urgent Shouting]. Poèmes. Forcauquier: Robert Morel, 1966.

Caves, Roger W. *Encyclopedia of the City.* New York and London: Routledge, 2004.

Crittenden, Jack, and Peter Levine. "Civic Education." *The Stanford Encyclopedia of Philosophy* (Fall 2018 Edition), Edward N. Zalta, ed. https://plato.stanford.edu/archives/fall2018/entries/civic-education/

Gurley, Jennifer. "Platonic Paideia." *Philosophy and Literature* 23, no. 2 (1999): 351–377. Project Muse. https://doi.org/10.1353/phl.1999.0040

Howatson, Margaret C, dir. *Dictionnaire de l'Antiquité: Mythologie, littérature, civilization* [Dictionary of Antiquity: Mythology, Literature, Civilization]. Paris: Robert Laffont, 1993.

Jäger Werner. *Paideia. Die Formung des griechischen Menschen* [The Formation of the Greek Human Being]. Bd. II., 19542. (19441). *Paideia. The Ideals of Greek Culture*. V. II. Oxf., 1947.

Jimenez, Marc. *La querelle de l'art contemporain* [The Quarrel of the Contemporary Art]. Paris: Gallimard, 2005.

Kant, Immanuel. *Kritik der reinen Vernunft* [Critique of Pure Reason]. Riga: Johann Friedrich Hartknoch, 1781, 1787, 1790. Reprint, Stuttgart: Reclam Verlag, 2006.

Kant, Immanuel. *Kritik der praktischen Vernunft* [Critique of Practical Reason]. Riga: Johann Friedrich Hartknoch, 1788. Reprint, Stuttgart: Reclam, 2008.

Lessing, Gotthold Ephraim. *Laokoon oder Über die Grenzen der Malerei und Poesie*. Produced by Delphine Lettau and Gutenberg Projekt-DE: https://www.gutenberg.org/ebooks/6889.html.images

Laocoon: Or, The Limits of Poetry and Painting. Trans. by William Ross. London: Ridgeway, 1836. https://archive.org/details/laokoon00lessuoft

Laokoon and How the Ancients Represented Death. London: G. Bell & Sons, 1914. https://archive.org/details/laokoonhowancien00lessiala

Morin, Edgar. *Oú va le monde ?* [Where Is the World Going?]. Paris: L'Herne, 2007.

Pappas, Nickolas. "Plato's Aesthetics." *The Stanford Encyclopedia of Philosophy* (Fall 2020 Edition), Edward N. Zalta, ed. https://plato.stanford.edu/archives/fall2020/entries/plato-aesthetics/

Rampnoux, René. *Histoire de la pensée occidentale: de Socrate à Sartre* [History of Western Thought: From Socrates to Sartre]. Paris: Ellipses, 2008.

Sartwell, Crispin. "Beauty." *The Stanford Encyclopedia of Philosophy* (Winter 2017), Edward N. Zalta, ed. https://plato.stanford.edu/archives/win2017/entries/beauty/

Vergil. *Bucolics, Aeneid, and Georgics of Vergil*, edited by J. B. Greenough. Boston, MA: Ginn & Co, 1900. Perseus Digital Library. http://data.perseus.org/texts/urn:cts:latinLit:phi0690.phi003.perseus-lat1

Vergil. *Aeneid*. Trans. by Theodore C. Williams. Boston, MA: Houghton Mifflin Co, 1910. Perseus Digital Library: http://data.perseus.org/citations/urn:cts:latinLit:phi0690.phi003.perseus-eng2:2.40-2.56

Winckelmann, Johann J. *Geschichte der Kunst des Altertums* [History of Ancient Art]. Dresden: 1764. Reprint, Darmstardt, 1982. Heidelberg historic literature. https://doi.org/10.11588/diglit.1343

4

DEMOCRATIZATION OF ART AND CULTURE

In contemporary visual culture, images highlight the objects and new environments, where physical and digital spaces coexist and interact in real time. If one compares the technology of ancient times with that of the present day, one finds that all of our senses have been adapted to other technologies and that perception has also changed. However, in measuring evolution and technical progress, the individual has always sought means of achievement through imitation in the arts, where poetic *mimesis* is the predominant technique, as presented in Chapter 3, "The Arts and Traces of the Ideal City." Consequently, to discern what is represented or not represented, one must possess the ability to evaluate the degree of image production. Here, some of the accomplishments of the precursor schools can be recalled, associating industry, arts and crafts, and later design. Therefore, this discussion, based on communication, art, and design, encompasses the limits of knowledge of the distinction between the real and the imagined as an issue susceptible to abstractions, which are guided by creation and technological evolution, cultural transformation, and socioeconomic development. These transformations have involved a large and complex definition of what art should be. Furthermore, they have occurred mainly through the constant new forms that characterize artistic achievements, bringing art closer to social and political reality with the aim of the "democratization of culture" (Benjamin 1939; Bourdieu 1998). Nevertheless, intriguing problems drive creativity and the development of techniques for overcoming the limits of language as representations of imagination, ideas, and thinking. This development involves the search for greater understanding, knowledge, and––above all––an enhanced capacity to communicate and relate within public life aimed at democracy. In this direction, the following questions arise: What is known about the meanings of visual and written languages? Which references in art history support social and technological development?

DOI: 10.4324/9781003314943-5

The technical evolution of writing and communication has been the main driver of civilization and the possession of knowledge. Without writing and communication, it would not have been possible to preserve or acquire knowledge, and we would not be present today, thinking about our legacy and the future of our species. In this sense, what possibilities exist for continuous democratic development through implementing creativity, techniques, and learning processes for problem-solving and improvement through the arts? On the one hand, creative practices consider resources and sociocultural stimuli, ranging from Johannes Gutenberg's work with the printing press to the 20th century and transition to the 21st century (Gopnik 2011). On the other hand, these publication practices are also related widely to art publications in their history of technical evolution, aesthetic and social reflections on the cultural context in Europe, and modern worldviews. In addition, the etymology of publication is considered: *publicacioun*, from the Middle French; *publicatio*, *publicare*, from *publicus*, which means public in Latin.

Thus, the analysis in this chapter is developed in three parts. The first part, titled "The Arts and Their Role and Pace in Urban Expansion," aims to situate the human condition in the origin of its communicative evolution. Since its very beginnings, the development of writing has been associated with the complex evolutionary capacity of the human species, as a result of a constant and intensifying communicative learning process. Based on historical sources, writing and visual language are presented to approach communication systems and technical evolutions as the main drivers of civilizations and knowledge. Thus, the successive social and cultural transformations caused by technological and scientific innovations are discussed. In this sense, the Arts and Craft and Art Nouveau movements are highlighted as precursors in urban space development. The Industrial Revolution forms the basis for this scenario of transformations in communication and the arts' social and economic role. Finally, the dissemination of information through publications is outlined to discuss the mass communication of increasingly urban and industrialized society, further stimulating industrial and technological development (Meggs and Purvis 2016).

In the second part of the analysis, titled "Designing the Public Space and the Rise of the Mass Society," the aim is to present the most critical social function of art, namely, to create a balance between humans and their environment. This balance is discussed by analyzing humans' worldview representations through technology. The consequences of this technology and the mass resources are considered the collective perception of the public appropriating forms of individual perception to configure the contemporary image in urban centers. Thus, it is supposed that the masses are responsible for all the new attitudes linked to the image, that is, the increase in participants' involvement (Hallin 2020). In the last part, titled "Fragments of the Urban Image: A Digital Option," an interdisciplinary approach is proposed to explore the possibility that perception remains a hypothesis for the meaning of contemporary visual culture, in its values and effects related to the fragments in connection and algorithmic rule in the imagined and concrete realizations of the urban image in motion (Grabbe 2018). Images and texts are developed, transmitted, and received from all areas, thereby conditioning us amid a polysemy of written and visual language. In this sense, knowledge is essential for the discernment of information. It is precisely on this basis that the democratization of art and culture related to the public space is discussed. Nonetheless, the novelty is carefully considered in this analysis into a constant updating and monitoring of all technological development, highlighting what is essential for human evolution.

The Arts and Their Role and Pace in Urban Expansion

Studies on the origins of humans have reported our ability to develop and adapt to the most adverse conditions in nature. The primordial need for the development of what we now know as societies and cultures began when our ancestors abandoned nomadic life for the cultivation of land, housing, and group living by building villages and communities. Such studies have focused on registries left by our ancestors from the Neolithic era (4,000 b.c.e.) and the first civilization, the Sumerians in Mesopotamia. Without these registries, we would have little knowledge of the beginning of civilization, that is, of the first inventions, techniques, and development in the formation of a community. Furthermore, there is nothing more essential to the formation of community than the evolution of human perception and cognitive capacities, which seek relationships and mutual understanding through the constant improvement of communication, thereby building common beliefs and habits. At first, Paleolithic humans communicated through sounds. The understanding and meanings attributed to those different sounds are what we understand as the origin of language and visual representations. This first visual representation is what we know as the origins of writing. Since its beginning, writing has evolved in association with the sophisticated evolutionary capacity of the human species with regard to language as a result of a constant and cumulative learning process. This capacity of the human species is related to many abilities that differentiate us from other animals. Among them, the abilities identified most commonly are memory, abstraction, imagination, thinking, reasoning, creativity, and problem-solving.

Next came the evolution of visual representation with geometric and abstract signs, points, squares, and other shapes related to animals. There are records of such paintings––pictographs—in the caves of Altamira in Spain and Lascaux in the South of France. In many regions of Africa, the Americas, and Asia, petroglyphs have also been found. These consist of figures recorded upon rocks, which are pictographs or ideographs, a symbolic language that represents ideas or concepts. These representations originated in both written and visual languages. In relation to writing, these representations are graphic depictions of the spoken language, which from Paleolithic humans onward progressed through a reductive process until reaching what we know today as the alphabet and its letters. This achievement greatly facilitated human communication, which over time gave rise to languages. Nevertheless, the visual language developed in Mesopotamia and Egypt, along with the hieroglyphs, papyri, and illustrated manuscripts produced by the Egyptians, represent a legacy of visual communication. Philip Meggs and Alston Purvis, in the book *History of Graphic Design* (2016), record these innovative achievements in Mesopotamia, which led to the development of the alphabet and graphic communication in the Phoenician and Greco-Roman world. We must also consider the East Asian languages[1] and all of its contributions to the world of the arts, techniques, and design. However, this study does not detail the rich history of the Middle Ages, of its illustrated manuscripts, the work performed by the copyists in the production of the flowing text, or the organization of the production of these manuscripts crafted in the *scriptorium*.

What matters most is the full development of the reproduction of art and publications on art from the Renaissance onward. That was the moment when artistic activity in the face of the desacralization that was occurring found another meaning in the world. Gradually, it engaged the interest of an audience that was slowly gaining access to books and various other printed materials. All of this was the result of printing techniques and their continuous development and improvement. One of the most important and utilized techniques both in the West and in Asia was xylography. This was a printing technique that used a wooden base with the relief of the content as an image––characters, ideograms, drawings, symbols, and texts––to be reproduced. In China, with a written language consisting of more than 5,000 characters, xylography was an essential technique. In the West, like written language, typography was one of the most significant inventions for the development of civilization. Most Western languages are written using approximately 24 letters, rendering the typographic technique the most appropriate for printing with its independent, mobile, and reusable types. This new technique facilitated not only reproduction and the wider distribution of printed matter but also predominantly a growth in the number of literate people and the transmission of information and knowledge. Moreover, the documentation of knowledge and techniques promoted discoveries and the development of science and technology.

The Gutenberg press, a printing technology with movable type, was invented in Germany, where xylographers, artists, and printing presses combined to develop illustrated books and pamphlets. The press then spread to Italy, the Netherlands, England, and France. Of course, we would not be here reading or enlightening ourselves, or reviewing or learning, if it had not been for the invention of writing, followed by typography and the Gutenberg press. Over the following centuries, printing technologies improved and contributed to the West's social, economic, political, and cultural development, ultimately resulting in the Industrial Revolution and––above all––the expansion of the capitalist process. All of these events in the arts were part of a period of political and social transformation that favored the new ideas and behavior of thinkers and artists. At that time, at the end of the 18th century, the bourgeoisie gradually gained in self-assertion, mainly with the advent of the French Revolution and the progressive conquest of the public space by commercial and industrial development. In this sense, another historical moment that would motivate technological development arose in the 19th century with the relationship between the arts, crafts, and industry, which gradually crept toward the Industrial Revolution, consumer culture, and local and global markets. The event in question, conceived by Henry Cole and Prince Albert, was The Great Exhibition of the Works of Industry of All Nations, which was staged within the Crystal Palace in Hyde Park, London (1851). Following this, further advances in printing processes produced significant results for the development of the printing industry. In addition to posters, brochures, and books, several journals and magazines emerged as the primary vehicle of communication.

A remarkable moment in this history occurred at the beginning of the 19th century in London, when the German Friedrich Koenig presented a project for the construction of a steam-driven printing machine using a single cylinder, which was later perfected using double cylinders. This machine was able to produce a large print run. At that time, 400 sheets per hour was considered a high rate for printing, which meant the immediate distribution of material to readers in a short time, that is, the transmission of news to a broader audience in less time. The machine was immediately purchased by John Walter of *The Times* in London. On November 28, 1814, *The Times* was printed using automatic, steam-powered printing presses, which was announced to readers within the paper. On December 3, *The Times* followed up with an article explaining the introduction of the new technology. A steam engine had replaced the mechanical machines and, consequently, the speed of production had increased. The printing machines contained up to four cylinders, and thus, the printing capacity per hour had doubled in a short time. Paper quality was improved to provide the primary substrate for printing. Thus, as the print run increased, the lower the unit cost became. Accordingly, the processes of the printing industry were optimized, making production cheaper (Meggs and Purvis 2016).

With the successive social and cultural changes that occurred in the 19th century due to technological and scientific innovations, the Arts and Crafts movement is highlighted in this analysis because of the representativeness and influence of the precursors of art publications, namely John Ruskin and William Morris. Ruskin was a British art critic and historian and the leading articulator, along with Morris, of the movement. According to Ruskin, in the middle of the Second Industrial Revolution, industrialization and technology advanced the gradual rift between art and society to a critical stage, thereby isolating the artist. The consequences of this Industrial Revolution directly affected the division of social classes and the exploitation of workers. Mass production was also introduced without any concern for aesthetic or functional quality. Moreover, this increased interest in the accumulation of capital through false and misleading advertisements for the provided products. The consequences were the pastiche of historical models and a decline in creativity and appreciation for design, which was executed by engineers without any aesthetic concern. Conversely, behind

Ruskin's theories was his belief that beautiful things were valuable and useful precisely because they were beautiful. Morris and Ruskin then founded the Century Guild, the goal of which was "to render all branches of art the sphere no longer of the tradesman, but of the artist." In addition, the arts of design were to be elevated to "their rightful place beside painting and sculpture" (Meggs and Purvis 2016). The guild developed projects imbued with the characteristics of Renaissance and Japanese design, which represented a significant moment in the Arts and Crafts movement by influencing the floral stylization of Art Nouveau. In addition, all of these ideas needed a medium of diffusion; however, in the absence of such a medium, the best approach was to create a space that would allow freedom for new ideas. Thus, *The Century Guild Hobby Horse* was born. *The Hobby Horse* was the first periodical in the 1880s to present the point of view of the English Arts and Crafts movement to the European public and to treat printing as a serious form of design. According to Meggs and Purvis (2016), in an article titled "On the Unity of Art" in the January 1887 edition of *The Hobby Horse*, Selwyn Image vehemently stated that all forms of visual expression deserve the status of art. He rebuked the Royal Academy of Arts for its very limited scope in representing art and design forms and thus recommended changing the name of the entity to the Royal Academy of Oil Painting. This conceptual innovation manifested basic ideas for design in its aesthetic and functional aspects, which were aimed at unification not only between arts and crafts but also between art and industry. Morris was also one of the precursors of relating technological and industrial development and the role of arts, design, and architecture to preserve the environment through natural resources, as he was aware of the long-term and irreversible effects of technological progress. Furthermore, Morris addressed the relationship of nature as the essence of humankind, art, and truth. Therefore, an art that separates the craftsman to meet an "intellectual" art marginalized in galleries and collections fails to appreciate the objects of everyday life or even decorative or ornamental beauty. Despite this, such an art does not cease to be mercantile. However, Morris argued that true art should be found in these functional household objects that belong to everyday urban life. Morris had an idea of progress involving real and imaginative clarity of vision, considering each person's visual experience for understanding. He had hoped with perceptual progress and posed in the arts a challenge to humanity's sense of nature, which by implying a return to nature contradicted the contemporary context of mass society and the predominance of an urban way of life. Morris foresaw the need for an inclusive and egalitarian art as a mediator between nature and civilization; however, he recognized the positive aspects of progress and science to recover beauty by striving for perfection. In this same sense, modern science and art began to produce artificial handmade objects. Through simulations, the products of human labor began to present inversions of the meaning of nature, labor, and art. This conception meets contemporary environmental concern as a challenge when considering Morris' ideas that urban space belongs to the whole public and that whoever voluntarily harms this property is a public enemy. In this sense, as previously discussed, the quality of urban life is considered a fundamental right and thus a fairer and more inclusive public space, where the arts also act for this purpose (Herbert 1992).

The initiatives of Morris and Ruskin were essential for the development of art democratization, primarily of the arts, which not only encompassed artistic aptitude, arts, crafts, and the evolution of printing techniques but also primarily discussed and substantiated a critique of the artistic and cultural universe of the time and published its results. In Western societies in the late 19th century, many of the poorer classes were beginning to read. Through reading, they sought to understand the changes in their social environment along with the innovations and knowledge derived from the Industrial Revolution and primarily from scientific discoveries. Subsequently, many popular kinds of literature appeared that were aimed at this population, who remained on the margins of progress. All of the wealth resulting from the Industrial Revolution could be turned into something tangible for those who lived on the margins, so that they might feel "included" in society. The second phase of the Industrial Revolution generated significant changes in the social and economic roles of communication and behaviors. Before the 19th century, the dissemination of information through books and pamphlets was the most dominant communication function. Given the pace and scale of urban expansion and the increasing level of mass communication, the industrialized society responded by rapidly expanding the number of printers of advertising material, posters, periodicals, newspapers, thereby further stimulating industrial and technological development.

What is essential is how human beings came to be in a civilized environment. First, as a result of writing and typography organizing the types into sequence and alignment, linear thinking and individualism were conditioned and literacy increased, which became increasingly prevalent in Western societies, along with an appreciation for registering information and archiving memory. The second factor concerns the creation of painting techniques with the use of perspective during the Renaissance. The perceptions of Western humans have developed since then, guided by the two visual systems of perspective and typography. Furthermore, these forms of perception still prevail in the advances made in communication technologies, even if they were potentially developed through the illusion of four dimensions and moving images. Today, according to Clay Shirky (2010), the author of *Cognitive Surplus: Creativity and Generosity in a Connected Age*, arguments about the behavior of current society deal with forms of collaboration, specifically contemporary society being democratized through its access to information: "The Gutenberg printing press produced the Reformation, which produced the Scientific Revolution, which produced the Enlightenment, which produced the Internet, each move more liberating than the one before" (Gopnik 2011). In general, the possibilities of communicative mechanisms as factors of change in society do not stem from technology itself, but rather from what humans do with it. There is no other way to relate except by communicating––a process achieved by humans through technology and media. In this sense, the arts, intrinsic to their techniques, can be considered precursors in the process of evolution and access to knowledge in societies, idealizing the spirit of the time.

Designing the Public Space and the Rise of the Mass Society

The transformations that occurred at the beginning of the 20th century were essential for the manifestation of the avant-garde and its influences on design. Amid these transformations, art and design were the paths that artists and intellectuals used to question values concerning tradition and innovation in society. Therefore, the objective linearity of social life no longer had the same meaning. These transformations were the consequences of the Industrial Revolution and a booming capital market. The art of the 18th and 19th centuries no longer represented the reality of this new pace of society for the European vanguard. Therefore, artistic and intellectual expression was divided between the conservatives and the modernists. Avant-garde ideas were always controversial and questioned the *status quo*. Through new representations of shapes and colors, they protested the social context. A series of modern movements such as Fauvism had a limited effect on graphic design in the 20th century. Others, such as Cubism and Futurism, Dada and Surrealism, De Stijl, Suprematism, Constructivism, and Expressionism, directly influenced the graphic language of visual communication within the century (Meggs and Purvis 2016).

During this period, through First World War and the Russian Revolution, creative innovation was relatively frequent due to the widespread devastation and social instability, when the need for the reformulation, reconstruction, and rethinking of society stimulated artistic and cultural production. Thus, throughout 1917, with the strong influence of the Arts and Crafts movement and Russian Constructivism, the Deutscher Werkbund (Association of Craftsmen) emerged in Germany. The Werkbund focused on the unification of arts and crafts with industry to elevate the functional and aesthetic qualities of production to offer quality, functionality, and low cost. In the same sense, the De Stijl movement also emerged in the Netherlands after being founded by Theo van Doesburg, associated with the artists Piet Mondrian, Bart Anthony van der Leck, and Vilmos Huszár and the architect Jacobus Johannes Pieter Oud among others. De Stijl defended the absorption of pure art by applied art by justifying that it would not be reduced to the level of the everyday object; rather, it would be raised to the level of art. With the De Stijl movement, the rationalization of design was conceptualized with an aesthetic of reduction that was marked in the Bauhaus in Weimar, Germany. This period saw the birth of the Bauhaus, an art school founded by Walter Gropius that focused on the interests of industrial development and social needs, paying attention to the functional and technical aspects. Then, in the passage from 1919 to 1920, from the housing culture and with the social principle of consolidating art to the people, Bauhaus influenced the way of life. The basis of an aesthetic and social theory of Bauhaus was characterized by the relationship between art, technology, and industry, exerting an influence—especially with the advent of design—from the early 20th century. In 1926, the Bauhaus was renamed Hochschule für Gestaltung (Superior School of Form), and the influential *Bauhaus Magazine* began to be published. This magazine and a series of 14 Bauhaus books (*Bauhausbücher*) became relevant to the dissemination of advanced ideas on art theory and its application to architecture and design. Kandinsky, Klee, Gropius, Mondrian, Moholy-Nagy, and Van Doesburg were the editors or authors of volumes within the series.

László Moholy-Nagy designed 12 books and explored photography in all its technical possibilities with regard to mastering light in frames using ordinary, everyday objects. Although the photogram technique was not new, the innovative aspect was the use of such a technique to derive formal results by means of light in abstract compositions. Moholy-Nagy, one of the Bauhaus artists, was alerted to the growing importance of the media from that moment on. He suggested to his colleagues at the Bauhaus that they consider creating artworks involving photography, as imagery works were gaining more space as a vehicle of communication in modern society. In other words, Moholy-Nagy was showing the Bauhaus artists the emergence of photography as a new communication tool that was being increasingly solicited in the messages and reports of the time. In principle, within the artistic universe, the modern period was essential due to the perpetual new forms that characterized the context, within which ideologies almost always defined the purposive content of the arts, bringing them closer to industry, the market, and the sociopolitical reality. To think about art, artists, freedom, and equality is also to understand that within the last 100 years, new forms of art and their techniques have been essential for cultural transformations that concerned the rupture of values, the constant search for solutions to social problems, and the changing of habits and behaviors. As a result of the Bauhaus lessons, any form of hierarchy that compared both art and design and outlined the differences between the artist and the craftsman would no longer be accepted by its members. Undoubtedly, many innovative results in the world of the arts and between artists were due to the freedom and equality experienced during the existence of the Bauhaus.

With reference to Pierre Francastel (1956), modern art, which is frequently imaginative, abstractly presents time and space and can participate in the daily life of the individual, through purely figurative signs as much as through artifacts. One can deduce, therefore, that even before language can evolve, arrive at writing, and reach an almost perfect capacity for communication, language—still pictorial—continues to unify the communication of different cultures, according to new forms of representation. Indeed, new technologies offer new possibilities, regardless of the similarities among images in a particular art at all times and in all places. Again, under the effects of modernity, Francastel (1956) inserts us into evolution by means of the temporal rhythm of life. For instance, in French society a century ago, recall that in 1850 in the work of Victor Hugo, the world traveled by wagon and spoke French, and that in 1950, Hugo himself in his work affirmed that the world would travel by plane and would draw and sculpt as in Paris. Today, one would say the world is moving toward being deterritorialized, and we no longer notice time or place. Morin (2007) posed the following question: Where is the world going? (*Où va le monde?*). One would question how humans can further configure our world as good and commonplace.

Drawing became technically reproducible through xylography, while the technical manager of the reproduction of writing was ultimately the press. Reproduction began with the first technique of xylography, followed by lithography, printing techniques, and copper plate etching. It was through employing these techniques that the graphic arts illustrated everyday life. At the end of the 19th century, with the appearance of photography and cinema, the process of reproducing images was at the same level as speech. Accordingly, the work of art has always been reproducible. The process of its technical reproduction has developed throughout history with great intensity, yet the techniques and means of reproduction have differed, along with the purpose of the reproduction and the relationships of the images as expressions, representations, narrations, or communications.

In Walter Benjamin's essay on technical reproduction ([1939] 2002), through his reflections, we can understand that technical reproduction does not permit the loss of a work's visual value in its moment of existence. Nevertheless, transformations of the original work over time permit the eventual loss of its content and authenticity, but not its visual value. This authenticity represents an origin, a material statement, and its testimony, which transmits a historical tradition as value. Through reproduction, everything is diffused, and tradition disappears because there is no longer any testimony of each of these historic moments. When the criteria of authenticity changed in the art production, all of the social functions of art changed. Art is no longer based on ritual but on politics, and that new situation has led art to acquire a new meaning. Further, the most critical social function of art has been to create a balance between humans and their environment, which is achieved in humans through their representation of the world through technology. The consequences of this technology and the mass resources available are the collective perception of the public appropriating forms of individual perception to configure the contemporary image in urban centers. So, the spectator's association of ideas is immediately interrupted by the change of image. Thus, the masses are responsible for all of the new attitudes linked to the image. Hence, the increase in participants has increased involvement.

Benjamin anticipated the secularization of art, the transformation of perception concerning new techniques of art reproduction, and the dissolution of the concept of art to the benefit of market strategies. Thus, some time ago, contemporary art became a system of management, planning, and economic profitability of cultural goods, which was proper of liberal democracy and post-industrial societies that had integrated art into their system of reproduction. This annulled any possibility of the arts as a critical medium, in a controversial and provocative role that is proper of artistic attitude. However, the image value—the artwork valorization—from Bourdieu's perspective does not depend on the artist, but rather on the cultural environment as a universe of beliefs when valuing the work of art as a fetish, which gives rise to the belief in the power of artistic creation—considering the work of art as the symbolic and valued object. It is not only up to the artists or writers, but also up to the complex system of belief production, the valorization of art in general, and the distinction of values under a whole structure of social relations involving commerce, industry, the media, governmental and nongovernmental institutions, teaching, and culture, among others. Bourdieu (1998, 376) described the interdependent system directing arts into traditional history, according to Benjamin, and fetishism in the name of the master, or when in the art social history, the analysis is limited to the context in which artists are inserted. Through the means of configuring the object, like the most traditional ones, artists present distinction and the valorization of their art, adding in their individuality their social origin and formation. Artists present the essence of the traditional model of creation in their work. Therefore, the artist becomes a leading creator of the valued artwork. Hence, Bourdieu added to Benjamin's reflections regarding the ambiguity of the artwork for the public or the public for the artwork with a question about the contribution in creating the value of the artwork and the creator.

Today, the urban environment is closely related to the digital production of images and the world market system, not only through the use and exchange values of public goods according to the relations of production and the productive forces, but also through the symbolic values that represent cultural signification. Both social and cultural history identify how to observe the transformations that have implied changes in the arts, culture, and all aspects of human achievement in its sociocultural representations in the configuration of the urban image. One of the most representative works of the 20th century that concerned this fragmented sense of the urban image is Benjamin's *Arcades Project*, which describes the years from 1927 until the end of his life (1940) and remains an incomplete work. This book describes the main characteristics of an image of a 19th-century capital––Paris—with its universal exhibitions, fashion, advertising, stores, passages, iron and glass constructions, streets, and more. The image fragments in a textual and visual narrative enable perceptions of an urban image in movement. One of the relevant aspects of these observations is the concept of "aura" developed by Benjamin in 1935. This concept is related to the reproduction process and effects of urban images—taking the image as art throughout its history, demonstrating the presence of the image through reproduction, and maintaining its characteristics of origin––that is, the time and place. In contemporaneity, the sense of this transference of values in front of the image is also seen as "a decline of aura" regarding the rhythm of the urban image and its distance from the origin, that is, traditional values. However, this decline is also interpreted as a revolution of all art and image functions, especially at the progressive end of tradition, which favors the immediate experiences of everyday life and is mediated by digital technologies. Initially, in his last version of *The Work of Art in the Age of Mechanical Reproduction* ([1939] 2002), Benjamin addressed the technical reproductions––in his time the press, photography, radio, and cinema––that would lead to the impairment of authentic experiences, thus clarifying that he did not believe that the decline of aura would enable the democratization of art and culture. He questioned whether the stars would favor the media as we know it today or whether they would be the ones favored. In addition to this classic reference, it is necessary to understand that the urban environment is about the appearances directly related to the images. In this sense, the status of the contemporary image is analyzed in its technical and polysemic nature, that is, in the diversity of its sociocultural meaning, considering the fragmented and constituent aspects of the visual system from analog to digital, given that "after the rise of the Internet and today's algorithmic rule, we are again on the verge of a significant paradigm shift" (Kaponen 2018).

Fragments of the Urban Image: A Digital Option

With the development of digital technology and network communication through the World Wide Web, a transformation occurred from analog formats of films, photographs, and printed images to digital formats––primarily to new productions created in digital formats. Thus, the characteristics of each form––film, photography, and television––and their unique values as media were maintained but only updated with the new digital technology and inserted onto the Internet (McLuhan 1964). The well-known technological convergence that occurs through the World Wide Web is currently the most significant result of the digital technological evolution. The concept of hypertext was central to the creation of the World Wide Web. Virtually all companies, organizations, and institutions use the Web for a diverse range of activities, including managing projects, offering distance learning, providing virtual tours of museums, ordering materials, and distributing company information. In this sense, graphical Web browsers allow pictures, sound, and video capabilities to be embedded in the text and applications to be embedded in the multimedia system. Multimedia is dependent on information and communication technologies for direct (nonlinear) access, interactivity, and integration with programs and applications. Its spatiotemporal nature is classified into three media forms, the respective characteristics of which are first, discrete, or static media such as texts, static images, and graphics. Second, there are continuous or dynamic media with content that changes over time, such as audio, video, and animation. Finally, there are immersive media, which present interactive information elements in 3D environments that allow substantial interactivity with the user. As for the degree of linear interaction, the user has no control over the multimedia reproduction process, which differs from nonlinear interaction, in which the user can interact through navigation. Information is accessed in a nonlinear and autonomous manner by the user through various means of representation and expressions of content. The user can interact freely with multimedia. The basic classes of multimedia applications and services are conversation, messaging, retrieval by accessing multimedia information servers, and broadcasting. It is, thus, understood that the digital production of images offers an interactive dynamic of fragments of the urban image. This production concerns the diversity of images transmitted in their forms, formats, and meanings to the visual culture. However, these several meanings are also related to the variety of the visual forms of perception with consideration given to how the visual experience is presented. In this regard, among the world of material things, the natural, traditional media and the virtual, technological transference occurs from the material, natural world, and from analog––which represents traditional media––to digital. In communication sciences, studies of communication technology by companies and academic institutions are increasingly concerned with the future of the human condition in its relation to virtual realities, as a process that transforms the "real life" of the individual through technology. This reference is based on debates at international conferences held by the University of Southern California Annenberg and published in the *International Journal of Communication*, founded by Manuel Castells and Larry Gross, who research new communication technologies.

Let us suppose that a tendency exists toward a virtual world or the transcendence of time and space. It would oppose the need for physical and self-demarcated spaces of territory or object appropriations; furthermore, "the concept of media systems would no longer be relevant to comparative research, or a national media system as a unit of analysis would no longer be relevant" (Hallin 2020). In its global condition, many theorists have also discussed the Internet as a way of transforming relationships, gradually interfering in several areas. The Internet is a phenomenon of enormous potential, because it allows people with little technical knowledge to access the network and participate in mass communication. The individuals are also perceived as the idea of a base, as the matrix of a series that will enter the production line of a democratization process under the illusion of a human being who is more conscious in their choices. An example is the cultural project Augmented Berlin, developed with augmented reality (AR), in commemoration of the 75th anniversary of the liberation of Germany and Europe from Nazism on May 8, 2020, and the End of Second World War. Augmented Berlin enables the user to interact with the scenery and historical context of the city of Berlin in retrospect until 1945. The urban environment, with its monuments, architecture, and urban objects, is transported to historical moments, placing users in each significant location and providing historical information (Wagner 2022). Through AR, the user "participates" in this historical moment. In this cultural project, the technology employed is AR, which consists of the user being present in a real environment—in this case the city of Berlin—and receiving through the Augmented Berlin app additional information through visual (re)productions (i.e., animations and 3D visualizations). The AR experience is made possible through a smartphone, tablet, head-up display, holography system, or AR glasses. In AR, it is possible to differentiate between the physical and the digital world, unlike in virtual reality (VR), which is only possible through VR glasses. The virtual world can be seen, heard, and felt. However, as stated by Grabbe (2018), "the full simulation of realistic textures that are fully adjusted to the functional dynamic of the skin and to the complex sensory dynamic of haptics is in an early stage of development"; therefore, some essential categories are required "because the perception of surfaces and textures cannot clearly be categorized in a single sensory structure." Therefore, on the contrary, they are "embedded in a very complex multisensory functional level." In addition, Grabbe described the possibility of image transformation, known as a *hyperaesthetic image*––*quasi-object* and *quasi-image*––that goes beyond the common visual experience and "enables a synchronization of *hapto-tactile* feedback, depicted image objects (in the moment of performance), sensory stimuli-processing and signification of meaning to stabilize a coherent user's perception by the potential of the medium" (Grabbe 2018). In addition, the diversity of audio-visuals, film production, advertising, and visual arts represents how images can construct the social imaginary.

In this context, visual fragments configure the images displayed in the public space by undergoing several configuration techniques, codifications, and optimizations when conveying virtual time and space in relation to the physical urban environment. One must consider society in all its dimensions. The digital world is just an extension as a constituent of the social imaginary. In this context of image transformations in full progress based on technological advances, it is urgent to research the empirical and theoretical confluence regarding perception for the development of new concepts and coherent analysis directives. However, the legacy of our knowledge has undoubtedly brought us to this stage, and therefore, the sequence must start from this foundation for evolution. Hence, transformation is the main characteristic of urban visual culture as a representation of a world conception, with characteristics and needs that vary according to culture, society, and economy. Creativity is presented in its realization processes as the meaning of the urban image in its current transformation process, while the reality of society is represented by the configurations of images. Thus, through visual culture, the action of the present develops around elements such as the economy, techniques, science, ideologies, and political will. These actions transform evolution, and within such a perspective, one can reflect on the conditions that allow innovation and creation in the face of the polysemy of images in this new sense, which are based on the media spreading the ideas and images of mass society. The copious production of images is possible in a context that is itself favored by the sociocultural dynamics. Finally, as an answer to the initial questions of this discussion about the public space and the democratization of art and culture in its communication process, some significant advances through new technologies, creativity in the arts, and sociopolitical actions were presented, which have contributed to the democratic urban development.

The configuration of the urban image in contemporaneity, for an aesthetic of everyday life, is verified through new interventions connected to communication and information technologies in synchrony with the urban rhythm, making images, and new dimensions of spaces possible. Political and social transformation is perceived mainly in public spaces. Moreover, contemporary cultural projects have enabled new forms of perception, mainly visual. This visual phenomenon presents a significant course in the history of civilization, social life, technological changes, and the individual's senses in their development in major cities. In the next chapter, "Visual Media Effects," an analysis is presented that aims, through a historical, iconological, and iconographic lens, to investigate and evaluate visual culture in its ethical and aesthetic effects found in the formal elements of art content.

Note

1 Raashid Hussain, Ahsen Raza, Imran Siddiqi, et al.: "The Chinese, Japanese and Korean, are the main East Asian languages. The writing systems of these languages partially or completely use the Chinese characters Hanzi, Kanji or Hanza." "A comprehensive survey of handwritten document benchmarks: structure, usage and evaluation." *EURASIP Journal on Image and Video Processing* 46 (2015): 1–24. https://doi.org/10.1186/s13640-015-0102-5.

Bibliography

Bardini, Thierry. *Bootstrapping: Douglas Engelbart, Coevolution, and the Origins of Personal Computing*. Stanford, CA: Stanford University Press, 2000.

Benjamin, Walter. "Das Kunstwerk im Zeitalter seiner technischen Reproduzierbarkeit" [The Work of Art in the Age of Mechanical Reproduction]. 1939. Reprint, *Medienästhetische Schriften*. Frankfurt am Main: Suhrkamp, 2002.

Benjamin, Walter. *The Arcades Project*. Trans. by Eiland, Howard and Kevin McLaughlin, prepared on the basis of the German volume edited by Rolf Tiedemann. Cambridge, MA and London: Belknap Press of Harvard University Press, 1999.

Bourdieu, Pierre. *Les règles de l'art* [The Rules of Art]. Paris: Éditions du Seuil, 1998.

Francastel, Pierre. *Art et technique aux XIXe et XXe siècles* [Art and Technology in the 19th and 20th Centuries]. Paris: Gallimard, 1956.

Gopnik, Adam. "The Information: How the Internet Gets Inside Is." *The New Yorker*, February 7, 2011. https://www.newyorker.com/magazine/2011/02/14/the-information

Grabbe, Lars Christian. "Image Transformation: The Hyperaisthesis of Digital Images." In *Yearbook of Moving Image Studies, Image Evolution: Technological Transformations of Visual Media Culture*, edited by Lars Christian Grabbe, Patrick Rupert-Kruse, and Norbert M. Schmitz, 78–99. Marburg: Büchner-Verlag, 2018.

Hallin, Daniel. "Comparative Media Studies in the Digital Age: Comparative Research, System Change, and the Complexity of Media Systems." *International Journal of Communication 14* (November 2020): 1–12. https://ijoc.org/index.php/ijoc/article/view/14550

Hussain, Raashid, Ahsen Raza, Imran Siddiqi, et al. "A Comprehensive Survey of Handwritten Document Benchmarks: Structure, Usage and Evaluation." *EURASIP Journal on Image and Video Processing*, 46 (2015): 1–24. https://doi.org/10.1186/s13640-015-0102-5

Johnson, Michael. "De Stijl (1917–1932)." In *The Routledge Encyclopedia of Modernism*. London and New York: Taylor and Francis, 2016. doi:10.4324/9781135000356-REM166-1

Kulturprojekte Berlin. "Augmented Berlin." 75th Anniversary of the End of WWII. Accessed August 8, 2021. https://www.kulturprojekte.berlin/en/projects/75th-anniversary-of-the-end-of-wwii/

McLuhan, Marshall. *Understanding Media: The Extensions of Man.* New York: McGraw-Hill, 1964.

Meggs, B. Philip, and Alston Purvis. *A History of Graphic Design.* Michigan: John Wiley & Sons, 2016.

Morin, Edgar. *Oú va le monde ?* [Where is the World Going?] Paris: Éditions de L'Herne, 2007.

Shirky, Clay. *Cognitive Surplus: Creativity and Generosity in a Connected Age.* New York: Penguin Press, 2010.

Wagner, Christiane. "Everyday Aesthetics and Augmented Reality: Political, Cultural, and Visual Contexts." In *Yearbook of Moving Image Studies: Augmented Images: Trilogy of Synthetic Realities II*, edited by Lars Christian Grabbe, Patrick Rupert-Kruse, and Norbert M. Schmitz, 17–37. Marburg: BüchnerVerlag, 2022.

5

VISUAL MEDIA EFFECTS

In global visual culture, recent representations and reproductions seem to present characteristics that appropriate the essential values of the human condition consecrated by classic Western art. In this sense, one should not necessarily consider that the sociopolitical meanings of images or actions are politically exercised. However, there are works and themes that allude to freedom of expression, ethics, and censorship. They appear in respective cultural events and general ways in the media, embodying the rhythm of the metropolises. Therefore, they present new elements that must be researched and discussed more systematically. Thus, this analysis concentrates on the absolute reference values of human existence in visual culture, eliminating any illusion, given the knowledge of novelty, in the face of mercantile injunctions and technical progress concerning its impact on international politics. Insomuch as ethics and aesthetics can be questioned, would it be the search for "the true in the configurations of images" in the sense of positive and determinate power or simply the practical consequence of vicissitudes in the formal diversity of creations? To provide insight on this preliminary problem, this discussion on the symbolic aspects of images focuses on the following main question: Can images prove facts transmitted by the media regarding the values of human rights? For this analysis to proceed, it is necessary to first ground this research in an iconology of representative images, such as in the relationship between photojournalism and classic artworks.

Answering these questions is the objective for addressing the main problem, especially concerning awareness of freedom and citizenship as well as respect for international norms and the ethics of human rights, as previously analyzed. Therefore, in "Visual Media Effects," an analysis based on the symbolic aspects of the elements that constitute democratic society's imagery is essential, along with reviews of conventional values of quality of life and the politicization of the individuals who represent it. Thus, iconology[1] becomes critical for this research, as visual culture

DOI: 10.4324/9781003314943-6

contributes to ethical factors through the perception of the individual—perception in terms of time-space and its social context (Panofsky 1955). Moreover, it is crucial to highlight visual culture as an experience of aesthetics and an evolution of its capacity for reasoning and cognition through the influence of signs and symbols in their social performance and autonomy (Bourdieu 1991).

The question regarding the true statement and transformative aspects of visual culture—in vogue—explicitly suggests a theoretical problem that exists in the challenge by thinking about the image—visual art in general—without merely inserting it into history. To complete a description and scientific analysis, in addition to contextualizing the artistic productions, techniques, and means that prevailed in their epochs, the discernment of ideological discourse is a fundamental prerogative. Such an epistemological commitment and methodological clarification aim to contextualize art in history when analyzing the current artistic dynamics. Moreover, the absolute reference values of human existence in visual culture are discussed in three parts in the face of its impact on international media.

First, in "Interpreting Contemporary Visual Arts and Photojournalism," the focus is on the critical exercise and understanding of both the iconological and practical structure that determines the symbolized image. Consequently, through aesthetic experience, an empirical analysis provides an understanding of the process through which it is possible to identify social facts in symbolic action through their function, thus understanding the meaning of the image's impact. The empirical element consists of the direct observation of the possible relationships presented and applied by media and public spaces—the media image—in major urban centers. Subsequently, in "The Visibility of True Values," the aim is to understand contemporary society through the perception, memory, or experience of the visual interfaces presented. It is primarily an analysis of the individual capable of creating new realities, determining the emergence of new forms of representation for political and cultural ideas linked to the media images of freedom and democratic values. In the third part, "Media and the Arts Regarding Children's World," the exploitation of images of children in the context of fundamental human rights is discussed, focusing on a critical approach to mass communication as well as the arts in its media functions. The need to recognize injustices and understand the developmental needs of children is evident (Piaget 1932, 1937, 1975). Nevertheless, there are still many goals to achieve to realize fundamental rights. Therefore, what is the ideal configuration of the image of children and its dissemination in the pursuit of these rights? The visual productions diffused by the United Nations Children's Fund (UNICEF)'s press image are one of the fundamental subjects of this analysis.

Interpreting Contemporary Visual Arts and Photojournalism

Art, photographs, films, videos, design, architecture, fashion, and more form part of society's complex system of visual culture. The use and understanding of visual images are governed by socially established symbolic codes. Visual images are constructed, and then they can be deconstructed and read as texts in various ways. Therefore, classifications are crucial for analyzing the impact and configuration of the social imaginary, wherein these images are produced and consumed. In this sense, the objective of this research is to analyze the visual products of society—their present production, consumption, and meaning. Images from different contexts and localities, with formal diversity and distinct contents, have generally exerted significant influence through the media in global and everyday social interactions, whether in economics, politics, science, or culture and its diversity.

However, regarding artistic achievements in the passage to the 21st century, there was a crisis and lack of meaning in the ideologies that were guiding modernity. Visual culture represents this trajectory, although without the ideologies that added a sense of sociopolitical participation and transformative image, subversive even to the system, without leaving aside the essence of this sociopolitical praxis, which is often provocative. If this type of art is understood in terms of Adornian aesthetics, it should not concern realistic content—that is, objectifying a critique of social reality with a specific subject and the intention of expressing a political ideology—because it simultaneously dissimulates an intentional propaganda. In addition, it is necessary to consider the ideologies that have given humanity a sense of sociopolitical participation as well as the transformative images that have subverted the rigidities of the system. In this direction, visual culture has presented consensually based democratic ideals and freedom of expression since the beginning of the 20th century. However, these forms of transgression have been transformed through a lack of ethical reference criteria and the dissolution of values. This explains why visual culture is often not understood or assimilated. In some cases, real tragedies are even confused with art because fiction so closely imitates reality. Take, for example, the global impact of the image of a photo taken by Burhan Ozbilici, a Turkish photographer from the Associated Press, which shows an armed police officer shouting and gesticulating, moments after fatally shooting the Russian ambassador to Turkey at an art gallery in Ankara. This photo resonated with people in the international media and was awarded "Photo of the Year"[2] at the 2017 World Press Photo Awards. Its selection for the award was not specifically about the violence itself but rather the photograph's composition. In addition, other images stand out for their exceptional composition that are not new but are already part of the sociocultural imaginary throughout the Western formation. Therefore, in a way, they condition the acceptance of their formal appreciation presented in different contexts.

It is challenging to find the true meaning of this image and to resist the influence of the power relationships resulting from the domination and injustice of globalized structures. As Bourdieu stated, "legitimate works thus exercise violence which protects them from the violence which would be needed if we were to perceive the expressive interest which they express only in forms which deny it" (1991, 139). Art has always depended on knowledge, learning, and the ability to transmit and express spontaneously or intentionally what is perceived in the environment. However, it is necessary to emphasize that art is the reproduction (*mimesis*) of balance, reflection, understanding, patience, happiness, joy, love, compassion, beauty, and goodness together with imbalance, irrationality, pain, unhappiness, desperation, and horror. Whether good or evil, we perceive these feelings in the visible universe and connect them with our personal experiences, as was already demonstrated in the third chapter "The Arts and Traces of the Ideal City." For this purpose, the results of this entanglement with artistic expressions grant creative productions and their symbolic values a greater dominion over our urban environment. Bourdieu (1991, 139–140) stated that, "specifically symbolic violence can only be exercised by the person who exercises it, and endured by the person who endures it, in a form which results in its misrecognition as such." Generally, political images—whether photographies, classic artworks, illustrations, or propaganda posters—act as instruments of analysis, producing emotional reactions and messages for the citizens of a society. Political images, in addition to generating responses to the central questions of this research concerning freedom, citizenship, and respect for international norms and the ethics of human rights, enable qualitative analysis, as opposed to quantitative data or interview transcripts (Bohnsack 2008).

In this case, media reality and society help to better understand the cultures concerning visual communication, that is, the effects of visual media and the perception of social reality (Luhmann 1996). The foundations of cultural production and reproduction allow access to systems of social structures, thus uncovering the motives and causes relevant to the functioning of the mechanisms of image production and reproduction. Since images in the context of sociocultural reproduction go beyond borders, it is necessary to enter into a reciprocal correspondence with the relationship between social structure, *habitus*, and *praxis*. The concept of *habitus* and *praxis* is a principle of the integration of reproductive practices into the social system. They refer to the aesthetic experience, perceptions, and learning processes as well as the interpretations, judgments, and behaviors conveyed and inculcated during the individual and collective unconscious through socialization and imagery production and reproduction. In *The Social Institution of Symbolic Power, Censorship and the Imposition of Form*, Bourdieu (1991) stated the following:

> Symbolic creations therefore owe their most specific properties to the social conditions of their production and, more precisely, to the position of the producer in the field of production, which governs, through various forms of mediation, not only the expressive interest, and the form and the force of the censorship which is imposed on it, but also the competence which allows this interest to be satisfied within the limits of these constraints.

Thus, the paradoxes of reality as transmitted by the media are seen by considering, on the one hand, that the individual can choose and select information consciously and, on the other hand, that the media can control and configure communication of the dominant reality. For example, the media and social networks can converge on a recent and essential phenomenon and give it a twist that essentially transforms it into "fake news." The concept of truth is complex, which is why there is much analysis of the difference between facts and opinions in the social sciences. Therefore, in terms of visual effects, we must be aware of analyses based on finding the fundamental visual elements for the true values of the human condition through artistic achievements within a social context. In its history, art has transformed the way we understand it—it does not matter precisely where technical evolution and the importance of scientific statute have contributed to elevating artistic production to a decisive stage.

The Visibility of "True Values"

True values, as absolute references of human existence, are an exploration of the eternal struggle for victory, justice, and freedom, through seeking to protect our liberty. They challenge humankind's fundamental ideologies at every level. As previously stated, the notion of freedom is philosophical, and therefore, through aesthetics we can apply it to artists and the artwork itself. Therefore, liberty can be considered the main concept and victory the secondary concept to position in regard to possible freedom, which is related to the dichotomies of true and reality through images associated with the eternal struggle for justice and freedom, maintaining the fundamental liberties at every level in humanity's history and, mainly, in the comprehension of our own evolution of how we relate to the notion of freedom.

An indirect analogy to Ozbilici's World Press Photo Award-winning image (2017) referenced in the previous section is the composition of visual elements in representations of the concepts of victory and freedom in the period of classical antiquity, specifically during the Hellenistic period,[3] in the image of the goddess Nike (Greek for "victory"),[4] the goddess Eleutheria (Greek for "freedom" or "liberty"), and the Roman goddess Libertas. In this sense, some works of art are representative of concepts that are essential to human existence. The evolution of these representations through a process of knowledge, reproduction, transmission, and interpretation has shaped Western culture, and they have provided *mimesis* of the essential values of humanity. The term *mimesis* refers to art as an imitation of nature and a representation of reality. Tragedy prevailed as the main conception of art for many centuries, which is one of the first formal examples of *mimesis*. From antiquity to the Middle Ages, the conception of reality and its representation was established, according to Erich Auerbach (1946), by The Odyssey and *The Bible*. While the real and the quotidian are often found in Homer's work, *The Bible* offers more interpretations. The use of daily facts as art represents signs of a superior advent, with symbolic and esoteric characteristics that are still appropriate for the medieval mentality. The Judeo-Christian religion and ethos were strongly related to everyday life and family, thus setting the stage for the sublime tragedy that marked the categorical rupture with antiquity in the sense of *mimesis* in its literary conception of reality.

Reflections on art in the West, up until Romanticism in the 19th century, were presented through interpretation of the theories of beauty and *mimesis*, as mentioned in Chapter 3, "The Arts and Traces of the Ideal City," finding renewal and a point of reference in the philosophy of Aristotle, as revealed in his *Poetics* (c. 335 b.c.e.), as embodying the highest achievement of tragedy and in contradiction to the ideas of Plato. In presenting *mimesis*, Aristotle demonstrated a refusal to separate the intelligible world from the sensible world, associating pleasure with the artistic imitation of nature. For Aristotle, art was beneficial in all its forms—for both the individual and society. For Plato, the sense of *mimesis* was to be found in the figurative arts, whereas for Aristotle, *mimesis* was a representation of feeling, equivalent in importance in artistic works to the meaning of *catharsis* in tragedy. Thus, employing iconography and its technical and material possibilities favors its expressive function. All of the recognized values of human experience and motivation were no longer related to the polytheistic culture of ancient Greece. In this specific case of the image, an example is the representation of *Nike* (goddess of victory) in an image of the Victory of Samothrace. Further, moving to biblical characters,[5] victory was transferred to the images of messengers of God or angels, such as that of Archangel Gabriel from the ad 6th to 7th century (Palagia Aggeloktisti monastery, Kitio, Cyprus); the artwork *The Annunciation* of Carlo Braccesco; and, during the Cinquecento, *The Annunciation* of Sandro Botticelli. In images from the Hellenistic period, the composition of the symbolic meaning of victory and freedom gained additional political and religious themes. An example of these features is found in the composition of Ozbilici's award-winning image from 2017. Here, we can draw an analogy to the composition of the painting *The Triumph of Samson* by Guido Reni (1612).[6] Reni's paintings primarily dealt with the themes of Bible stories and Greek mythology. According to the historian Klaas Spronk (2014), *The Triumph of Samson* can be seen as a mixture of mythological scenes characterized by a focus on the proportions of the human body. Reni added the visual references of Greek antiquity to the symbolic values of the Romans to create images that were representative of the experience of victory and of freedom consequently transferred to biblical characters. The reference used by Reni was the statue of the Apollo of Belvedere (currently in the Vatican Museum), which underscores the reference to concern for the anatomic representation and image of the human body in its balance of proportions.

For any representation of the concepts of victory and freedom, any attempts at verbal description become just as complex. Thus, how should victory and liberty be represented? Many of the images that have searched for these representations since classical antiquity have created a personification of "freedom" and "victory" as female characters related to patriotic themes. In ancient Greece there was the goddess of liberty, *Eleutheria*, while for the Romans it was *Libertas*. In France, freedom is personified by Marianne and in England by Britannia, while in the United States the figure of Columbia is the hallmark of liberty. The configuration of these personifications maintains similar attitude expressions of freedom characterized by symbols and body postures. Likewise, they are associated with the characteristics of victory found in the image of the Greek goddess *Nike* and the Roman goddess *Victoria*.

The history of art and literature has presented effective modes of influencing the configuration of images that impose the perception of their meanings and methods of structural or semiological analyses. Therefore, art as a symbolic commodity exists as such only for those who hold the means to decipher and appropriate it (Bourdieu 1991). An image remains subject to a variety of interpretations, especially when presented without textual references, because we first realize it as an aesthetic experience before we move on to an understanding of the full meaning of the image. In this way, one considers that the definitions are restricted to a specific realm of cognition or common sense, due to judgments based on subjective values being significant in terms of the polysemy of the image, especially given an immense diversity of values for image interpretation. In *Symbolic Power and the Political Field*, Bourdieu (1991) stated that the instruments of knowledge and communication are "a structuring power only because they are structured." He went on to explain symbolic power as follows:

> Symbolic power is a power of constructing reality, and one which tends to establish a gnoseological order: the immediate meaning of the world (and in particular of the social world) depends on what Durkheim calls logical conformism, that is, "a homogeneous conception of time, space, number and cause, one which makes it possible for different intellects to reach agreement."
>
> *(Bourdieu 1991, 166)*

By analyzing the current artistic dynamics through aesthetics and iconology and contextualizing the image's production, techniques, and means that prevailed in the epoch it was created in, the discernment of ideological discourse becomes fundamental to reaching a significant answer regarding how the feeling of freedom arises and encourages respect for human rights and fundamental freedoms. This is what the revolution was about—to leave the old for the new. This is the motivation and momentum for innovation, change, and victory. A recent sample of a democratic society's imagery is *American Uprising*, an illustration by Kadir Nelson for *Rolling Stone* magazine. It is analogous to *Liberty Leading the People* (*La Liberté guidant le peuple*) by Eugène Delacroix (1830), which depicts a woman leading the French Revolution. In his interview with *Rolling Stone*, Nelson stated, "I wanted to create something in a similar vein that would have the same feeling and celebrate this tipping point in our history." He continued that "the people who were pushing for those changes were African American women" (Aleah and Martoccio 2020).

Media and the Arts Regarding Children's World

Here, the visible aspects of the perception and evolution of childhood are analyzed through art and images concerning the fundamental rights of children and the representation of these images in the sociocultural context. Hence, images of children associated with social imagination are also highlighted through art and media. In addition to artistic and photographic representations, the materialization of the child's image is observed through the symbolic aspects of visualization and interpretation. It must also be considered that a universe of imagination is always present. The images constructed in our thoughts are conditioned by our experiences and senses, which are, in turn, influenced by our training and cultural knowledge. Thus, two aspects are described. The first is the relationship of the image of the child in the public space, and the second is the image of the child associated with the social imagination through art and media, mainly visual productions that incorporate representations of children disseminated in the media by UNICEF. In this regard, there is an evident need to recognize injustices and understand the developmental needs of children. To realize the ideals of human rights, there are still many goals left to achieve.

Following in the path of Philippe Ariès (1973) in his analysis of medieval art, childhood was not represented until the 12th century as sacred art predominated. However, in these medieval paintings, when a child was depicted, he or she was portrayed with the characteristics of innocence associated with the nude, and the body and face were those of adults, but in a reduced size. The child most represented in sacred paintings was the child Jesus, who was always linked to the maternal figure of the Virgin Mary. The other images of children are of angels and saints. These sacred paintings aimed to propagate Christianity and predominated until Romanticism, glorifying the absolute monarchs and the taste of the bourgeoisie. Gradually, religious iconography was transferred to a secular and bourgeois iconography of childhood.

During the Renaissance, in a scenario where humanism and new ways of conceiving life in society prevailed, the meaning of childhood also became a point of reference and sensitive representation in art. In the paintings of Andrea Mantegna, Tiziano Vecellio, and Raffaello Sanzio, images of children are represented within a framework of the conceptions of Christianity. However, the maternal image related to the care of the child by the Madonna is an image of medieval art that has persisted until today. Technically, Renaissance paintings represent a significant plastic evolution: Images of adults with reduced proportions were no longer used to represent children. Instead, the compositions included images of children with their characteristics playing in social spaces, and these spontaneous movements were well reproduced. These characteristics developed in the Renaissance extended into Romanticism. Single child portraits appeared in the 17th and 18th centuries, as did family and memorial pictures featuring images of deceased children. The iconography of the dead child was representative of the sense of rightness—piety—that was linked to their fragility. From the 19th century, secularization started to increase, which has continued until today. As a result, new forms of representation of the child's world have been associated with the social history of children in terms of development and identification by society. In general, before the Renaissance, the child as a social subject was only a character in the composition of illustrations of festivals in medieval art. Instead, children played around their mother, and other characters gathered in public spaces. In Pieter Bruegel's painting, *Children's Games* (1560), this iconographic sense can be perceived as the reality of medieval public spaces with cultural festivals, the circus, street theater, and everyday life. Thus, how medieval people related to children can be observed. However, let us consider the history of games and attractions in public spaces throughout the evolution of urban life. It is only recently, in the 1920s, that the playground emerged in public spaces in the way we know it today. It introduced specific and functional toys encouraged by the demand of the mass population to meet new economic and social needs in large cities. Notably, this social reality became the image of childhood represented in many contemporary media. As an example, Bruegel's painting is only the first achievement of this representation of the social imaginary of childhood. Therefore, this reality of urban and representative life in paintings is due to the context in which children's rights become valued. As playgrounds have only recently become truly commonplace, they are considered mere by-products of the 20th-century city. These children's spaces were created within the confines of the city to meet the needs of the working class, whose children found themselves without parental care during the day; they served as lively places in urban societies. This same movement of creating children's play spaces has emerged in Brazil.

In 1920, Fernando de Azevedo initiated directed recreation programs characterized by public parks, which resulted in the first playground in São Paulo in the early 1930s. In 1935, playgrounds (with built-up areas and open spaces) were implemented, following the American Reform Park model. The modernist Mário de Andrade, in the newly created Department of Culture, was the lead designer and had the ideal of promoting moral and cultural assistance to the youth and children of the poorest classes. These new values in childhood are part of a social construction based on public participation. The different forms of visual representation form the social imaginary. However, this analysis focuses on the child's image as a subject and not as an object. As an image, the bivalence of the child is both object and subject in the historical context of the development of Western societies. These images of children are an integral part of the historical moment and the subject from the moment the child is socially recognized as having the right to a childhood made of development, learning, and formation. Therefore, in this transitional phase from the Renaissance to the modern era, the child's image was linked to the stereotyped and idealized reality of the family by the nobility and bourgeoisie of traditional European societies. In Brazil, during colonization (1500–1815), the reality idealized by traditional European societies exerted a strong influence. However, it was not until the beginning of the 20th century that the first images describing children's social reality in the urban and rural context, through successive modernist movements in which the arts revolutionized the representation of the child, were found.

Furthermore, while sociological studies concerning childhood only became significant in the 1990s, the artistic scenario had already found this meaning at the beginning of the 20th century, when art definitively moved away from the canons of fine art—that is, from the ideals of beauty and the divine—to confront the social and political reality. At this point, the inhuman aspects of the effects of the First World War represented by art appeared. The fundamental social problems were industrialization, maternal life, and child labor, which were represented by expressionist art and the New Objectivity (Neue Sachlichkeit) work of Otto Dix in Germany. The art of the New Objectivity had as its "objective," in the strict sense of the term, to denounce the problems of this post-war society by using figurative images to highlight human misery as a critique of bourgeois society. Would it not be the purpose of the media to explore images of children in miserable, desperate situations and conditions of suffering? We see this visual representation of children reproduced with great intensity in the media today. However, this analysis contests this same condition by asking the following question: Is it appropriate to show politically correct visuals of children or images of vulnerable situations?

In Brazilian culture, the representative image of childhood in the works of Candido Portinari is of great interest, as he not only depicted the reality of Brazilian society and culture as well as the relationship of childhood in this context as a time of toys, games, and happiness, but also included details that represented moments of human tragedy. Portinari's work is of great importance precisely because of its strong appeal in illustrating a society and culture searching for its values and rights and its development and formation as a former colony of Portugal. From an era represented by the arts and the press to today's digital media, these images represent the construction of the social imaginary and the formation of the current visual culture. Even before the appearance of the United Nations and, recently, the Convention on the Rights of the Child, we can see that the representation of childhood values is not only Brazilian but also universal. In this perspective, aesthetic analysis finds the conceptual formula of universality in its classical Kantian and Hegelian principles. These values are beyond the immediate and objective experience; in addition, they are fundamental and proper to the human condition in its existence, of which the stage of life, namely childhood, is a part. In this sense, through an empirical qualitative analysis, references to symbols, meanings, and signifiers are identified in the contemporary images of UNICEF, which indicate the traditional values present in a privileged societal class that maintain and visually reconfigure a universal childhood.

Through comparing Portinari's artworks depicting Brazilian childhood with the current images of UNICEF Brazil, the analysis seeks to address the following question: Are new conceptions of images possible or do they follow a representation of traditional social imagery, as in Portinari's works? The image configurations highlight how the social imaginary maintains tradition in its visual representation. Of course, the context is different, but the image of childhood remains idealized while ceasing to present the social reality. In this case, the analysis concerns the representativeness of Portinari's artworks *Black Mother* (1940) and *Child with a Kite* (1947) with UNICEF's media images, among other images representative of the playful imagination of childhood in Brazilian culture that reveal UNICEF's media functions and Portinari's collection of works. These images have been selected to trace the image's meaning through the child's social history through art and its representativeness in the social construction of childhood values through the media, which must be respected as fundamental rights. Thus, throughout the formation of Western society, in the face of the polysemy of images, the playful image is highlighted, exploring the aspects of ingenuity, joy, and fun as imaginary of the social construction of childhood. However, through some of these charitable institutions, we still find today the exploitation of the child's most undignified and desperate conditions with images captured in moments of terror, fear, and suffering. According to the Convention on the Rights of the Child, the dignity of every child must be ensured and respected. Therefore, in this way, UNICEF follows strict guidelines for photography, which ensures that the dignity of every child is respected. Thus, if the preservation of the dignity of and respect for the child are fundamental rights, how can we claim to be preserving these rights by exposing situations of vulnerability, mistreatment, poverty, hunger, suffering, and destruction? Only one thing is essential, and that is to help protect these children and ensure that their rights are respected. For this, a piece of paper symbolizing this imperative—the Convention on the Rights of the Child—is not enough. Still, it is necessary to enforce the provisions of this precious document with all the political goodwill possible and, undoubtedly, to raise awareness in society. For this, it is essential that children are respected and that their image is protected. When we talk about an image, we are talking about the concrete image and the one we refer to by memory, associating it with a name or a situation. In this sense, we must know the limits of the exposure of photos and videos in the various media, such as social media and the press in general, or through the theater, cinema, and various cultural events. Further, the conditions of exposure of images of children must be exceptionally well regulated in the context of advertising appeals of some nongovernmental organizations or institutions dedicated to children's issues, which is a common practice. In line with this provision, in 2016, children from a school in South London asked some of these organizations to stop using emotional images of children in appeals. This exposure to children's images is due to the various advertising campaigns that involve an exploration of tragedy in the world of children.

The children in this South London school were aware of the Convention on the Rights of the Child and the fundamental role of UNICEF. However, as a means of denouncing social problems, was the exploitation of images of children in miserable and desperate situations and conditions of suffering not the goal of the media? What must be understood is that, in fact, and under law, the images of these children in vulnerable situations must be preserved. At the same time, it is essential to think about measures and strategies for making society, and more specifically the social elite, aware that there is indeed a dark and decaying world outside of their privileges and well-being. Suppose only the image of an enchanted world prevails, of well-fed, smiling children on their parents' laps, with toys in their hands or playing in amusement parks or their schools. In that case, it would not mean that these images are a unique reality. In most situations, they only express a desire. What exists is precisely an illusion of happiness. For the social elite to become aware of inequality and social injustice and for the rights of the child to prevail, the best strategy is to ensure access to education, literature, and art. Transmitting culture means transmitting what has already been learned by others so that future generations do not need to start new processes but rather build on existing ones, experimenting with the mistakes and successes of the past for new possibilities.

Nevertheless, with the development of technologies, transmission functions have changed, and the mediations between people have taken on new meanings. The processes have become increasingly complex as they involve new interactions, exposures, and contexts, a set of signs, symbols, and signals, with the participation of individuals. The media and social transformations affect and accelerate the cultural process and facilitate communication between the different layers of society. Thus, these cultural and social mediations have impacted literary works in the function of consumer goods. From the second half of the 20th century onward, the media's presence and behavior in society have transformed the mass of individuals into subjects of information, where the focus of educational objectives is to prepare the child for the construction of knowledge (Piaget 1975). The media is also seen as a new organization in the face of these transformations of the informed society, through generalized intelligence and the relationship with institutional and universal bodies, such as the United Nations Educational, Scientific and Cultural Organization (UNESCO) and UNICEF, which interfere in international politics, education, media, industry, and the market.

The world's visual culture has the distinctive features of images depicting social inequality and, indeed, human representations—that is, the essential aspects of the human condition enshrined in classical, modern, and contemporary art, film, and media. Moreover, most social themes allude to human rights, demonstrating the fight against inequality through visual representations in the real world. Therefore, the sociopolitical meanings of images used in a political context regarding ethics and aesthetics are not necessarily considered. Hence, the search for politicization in image production or the vicissitudes of practical consequence in the formal diversity of image configurations has been confirmed—as far as the images analyzed are concerned—regarding the representativeness of classic Western paintings with the media images. However, to elucidate these observations, the analysis was discussed through the symbolic aspects of the images following two forms of investigation: First, the exploitation of tragedy as a common motif of visual experiences in the media was analyzed, and second, the "real" symbolic values were proved to be factual through the images based on the impact of inequalities in visual representations. Thus, a visual analysis based on the symbolic aspects of the elements that constitute the imagery of democratic society as well as a review of conventional values regarding the awareness of fundamental human rights were essential.

In this sense, the objective of "Visual Media Effects" was to analyze the image in contemporary society and its production, consumption, and meaning. The results were based on the discovery of fundamental visual elements—observing social inequality, justice, and freedom as facts through iconography, iconology, and aesthetics—to understand the representativeness of the human condition in the media and artistic achievements. The values of this human condition were thus presented, eliminating potential illusions through aesthetic experiences, analytical logic, and the possibilities of visual representations. The next chapter, "Urban Resilience: Toward Postcolonial Aesthetics," analyzes current media images and presents essential values of significant artworks that reflect the sociocultural context both inside and outside of Europe as well as the consequences of the reference systems in postcolonial cultures.

Notes

1 Paul Taylor:

> The words "iconology" and "iconography" are often confused, and they have never been given definitions accepted by all iconographers and iconologists. Panofsky 1955 (cited under General Overviews) defined "iconography" as the study of subject matter in the visual arts and "iconology" as an attempt to analyze the significant subject of that matter within the culture that produced it. This definition was prescriptive rather than descriptive, and many art historians before Panofsky, who would have called themselves "iconographers," were, in fact, engaged in investigations that Panofsky would have termed "iconological.
>
> *Iconology and Iconography* (Oxford Bibliographies, 2017), doi: 10.1093/OBO/9780195399301-0161

2 Diane Smyth: "The story, and particularly the World Press Photo-winning image [2017], quickly went viral, but Ozbilici says he didn't think about 'how people might distribute it in social media, or how they might use it.' 'I didn't pay any attention to that, and I was trying to do my work,'" he says.

"The assassination was, of course, political, but I am not about this, I am only a journalist. I am trying to record, to document, to witness history." In "Burhan Ozbilici wins the World Press Photo of the Year." *British Journal of Photography*, 13 February 2017, accessed May 12, 2022, https://www.worldpressphoto.org/news/2017-02-13/world-press-photo-announces-winners-2017-contests.

3 Louvre Museum:

> The Greeks represented concepts such as Peace, Fortune, Vengeance, and Justice as goddesses at a very early date. The victory was one of the earliest of these incarnations. She is a female figure with large wings that enable her to fly over the earth spreading news of victory, whether in athletic competition or battle. She is a messenger (Angelos in Greek) who sometimes uses a trumpet to make her message better heard. As she flies, she brings the victor the symbol of victory—a crown, fillet, palm, trophy of arms, or naval trophy. . . . Victory is an extremely decorative figure who appeared widely in Greek art from the Archaic period (sixth century BC) onwards. She is found in a multiplicity of forms—statues, reliefs, vessels, coins, and terracotta or bronze figurines. Such figures followed the stylistic evolution of Greek art, undergoing constant development. As the Victory of Samothrace shows, the figure still featured in spectacular works of art in the Hellenistic period.
>
> Accessed May 8, 2022, http://musee.louvre.fr/oal/victoiredesamothrace/victoiredesamothrace_acc_en.html#seq_8

4 *Nike of Samothrace*, Musée du Louvre, Paris, Greek original by the sculptor of Rhodes derived from type C4th B.C. Winged Victory of Samothrace circa 220–185 B.C., accessed April 9, 2022, http://musee.louvre.fr/oal/victoiredesamothrace/victoiredesamothrace_acc_en.html.
5 Louvre Museum:

> Angels holding globes and crosses stood close to God as representations of his power and glory. However, although angels owed part of their role to Greek and Roman representations of Victory, their image was rather different. Early Christian depictions of angels show them with a halo and dressed in male garb typical of Antiquity—a long, wide-sleeved tunic covered with a pallium, or long cloak, worn draped diagonally across the chest or thrown over the shoulders. They were winged messengers who came down from heaven to announce God's will to mankind. Angels only began to wear female garments in the late medieval period, when the draped cloak was no more than a small drapery worn like a shawl and the tunic became an elegant tight-sleeved gown with a high waist. The artistic popularity of antique models during the Italian *Quattrocento* meant that angels began to resemble female Victories, although the Christian context leaves no doubt as to their identity.
>
> Accessed April 15, 2022, http://musee.louvre.fr/oal/victoiredesamothrace/victoiredesamothrace_acc_en.html#seq_8

6 Pinacoteca Nazionale di Bologna—Room 24:

> Samson is the highest expression of that idea of balance, beauty, and harmony that inspired Guido Reni. To portray him, the painter studied a pose that draws on the classical model of the statue of the *Apollo of Belvedere* today at the Vatican Museums. The filter of ancient beauty annuls all traces of suffering and fatigue over the struggle he has just undertaken. Instead, there is a greater naturalism in the bodies of the Philistines, piled on the ground, bloody and tragic, left in the shade, sharply contrasting with the luminous quality of the victorious Samson.
>
> Accessed April 18, 2022, https://www.pinacotecabologna.beniculturali.it/en/content_page/item/2823-samson

Bibliography

Adorno, Theodor. *Ästhetische Theorie* [Aesthetic Theory]. Frankfurt am Main: Suhrkamp, 1970.

Aleah, Kimberly, and Angie Martoccio. "Kadir Nelson's Inspiration for the July Cover of *Rolling Stone*." *Rolling Stone*, June 16, 2020. https://www.rollingstone.com/culture/culture-news/kadir-nelson-cover-artwork-july-2020-1014011/

Ariès, Philippe. *L'enfant et la vie familiale sous l'Ancien Régime* [The Child and Family Life Under the Ancient Regime]. Paris: Seuil, 1973.

Auerbach, Erich. *Mimesis: dargestellte Wirklichkeit in der abendländischen Literatur.* 1946. Reprint, Basel: Francke, 1994. *Mimesis: The Representation of Reality in Western Literature.* Trans. by Willard R. Trask. Princeton, NJ and Oxford: Princeton University Press, 2003. *Literatur.* Bern: Francke.

Bohnsack, Ralf. "The Interpretation of Pictures and the Documentary Method" [64 paragraphs]. *Forum Qualitative Sozialforschung/Forum: Qualitative Social Research* 9, no. 3, Art. 26 (2008). http://nbnresolving.de/urn:nbn:de:0114fqs0803267

Bourdieu, Pierre. *Language and Symbolic Power.* Cambridge: Polity Press, 1991.

Burocco, Laura. "Reflections on the Future and Past of Decolonisation: Africa and Latin America." Third Text, Critical Perspectives On Contemporary Art and Culture. September 1, 2021. http://thirdtext.org/burocco-reflections

Hegel, Georg Wilhelm Friedrich. *The Phenomenology of Spirit.* Trans. by Terry Pinkard (Cambridge Hegel Translations). Cambridge: University Press, 2018. Originally published as *Die Phänomenologie des Geistes.* (Bamberg and Würzburg: Joseph Anton Goebhardt, 1807).

Hegel, Georg Wilhelm Friedrich. *Lectures on the Philosophy of Art.* The Hotho Transcript of the 1823 Berlin Lectures. Trans. by R.F. Brown. Oxford: Clarendon Press, 2014. Originally published as *Vorlesung über die Philosophie der Kunst. Berlin 1823.* Rewritten by Heinrich Gustav Hotho (Hamburg: Felix Meiner, 1998).

Kant, Immanuel. *Kritik der reinen Vernunft* [Critique of Pure Reason]. Riga: Johann Friedrich Hartknoch, 1781, 1787, 1790. Reprint, Stuttgart: Reclam Verlag, 2006.

Kant, Immanuel. *Kritik der praktischen Vernunft* [Critique of Practical Reason]. Riga: Johann Friedrich Hartknoch, 1788. Reprint, Stuttgart: Reclam, 2008.

Luhmann, Niklas. *Die Realität der Massenmedien* [The Reality of the Mass Media]. Wiesbaden: Westdeutscher Verlag, 1996.

Panofsky, Erwin. *Meaning in the Visual Art: Papers in and on Art History.* New York: Doubleday and Company, 1955.

Piaget, Jean. *Le jugement moral chez l'enfant* [Moral Judgment in Children]. Bibliothèque philosophie contemporaine. Paris: PUF, 1932.

Piaget, Jean. *La construction du réel chez l'enfant* [The Construction of Reality in Children]. Paris: Delachaux et Niestlé, 1937.

Piaget, Jean. *L'équilibration des structures cognitives: Problème central du développement* [Balancing Cognitive Structures: A Central Problem in Development]. Paris: PUF, 1975.

Rousseau, Jean-Jacques. *Du contrat social ou principes du droit politique* [Of the Social Contract or Principles of Political Law]. Paris: Classiques Garnier Frères, 1960.

Spronk, Klaas. "The Looks of a Hero: Some Aspects of Samson in Fine Arts." In *Samson: Hero or Fool? The Many Faces of Samson*, edited by Erik M. M. Eynikel and Tobias Nicklas, 197–209. Leiden: Brill, 2014.

Wagner, Christiane. "Artworks and the Paradoxes of the Media-Transmitted Reality." *Conference Proceedings, International Congress of Aesthetics, AM Journal of Art and Media Studies* (October 2019): 71–85. http://doi.org/10.25038/am.v0i20.324

6

URBAN RESILIENCE

Toward Postcolonial Aesthetics

In the previous chapters, notions of the Western worldview's construction have been presented, addressing the meanings of truth, reality, freedom, conscience, fundamental rights, and, in general, the construction of an aesthetic knowledge with French, German, and Anglo-American origins. However, it is necessary to consider some subtleties in the contemporary context involving aesthetics when the object is art engaged in political practice as part of theoretical criticism. After all, on the one side there is the commitment of activist artists, while on the other side academic work in the human sciences seeks the fundamentals and reality of the facts in studies of social and cultural phenomena through methodologies compatible with the research at hand. In this sense, this chapter focuses on postcolonial aesthetics to discuss theoretical challenges, seeking to overcome the generalized notions of modern European thought through an investigation of non-Western nations that seek a break with hegemonic Western forms of globalization. The postcolonial approach involves all knowledge, including aesthetics—and by extension the arts—featuring cultural diversity and recognizing otherness in various artistic forms. These recent postcolonial theories have discussed the production of knowledge by considering a structured set of concepts, assumptions, and discursive practices, which have been used to produce, interpret, and evaluate knowledge about non-European peoples (Chakrabarty 2000; Mbembe 2001; Mignolo and Walsh 2018; Appadurai 2021).

Consequently, these theories aim to deconstruct Western worldviews as part of the imperialist project, relating power and knowledge not only to colonialism's economic or political logic. The discussion in this chapter is guided by the following questions: What aesthetic values are considered essential without being compared between the South and the global North? Is it possible to understand the concept of beauty in its deeper qualities of the human condition that overcome the notions of modern European thought? Could the arts, in the face of

DOI: 10.4324/9781003314943-7

globalization, with aesthetic values related to the concept of identity, offer education through a positive or negative utopia? Can the arts overcome capitalist and imperialist political-economic systems and patriarchy through the global South's arts for a more egalitarian society? To answer these questions, sociocultural, political, ethical, and aesthetic values are discussed, especially the cultural values of native peoples and Afro-descendants in contemporary times. Therefore, certain factors are considered, such as aesthetic appreciation and the notion of artistic beauty and its reverse. This analysis focuses on the current context, considering cultural differences, especially the global South and its indigenous and tribal peoples' cosmologies, Afropolitanism, and the Afro-diasporic movement.

Thus, postcolonial aesthetics is considered a process of transformation, resistance, and positive and negative utopias in search of overcoming colonialism. In the first section, "Overcoming (Post)Colonialism," the notion of postcolonial aesthetics is discussed along with the fundamentals of aesthetics as part of the construction of modern European knowledge and its contemporary paradoxes for decolonization with new worldviews. Latin America is on one side and Africa is on the other. In the second section, "Epistemological Rootedness and Aesthetics Toward Decolonization," the development of modern European knowledge is discussed with a focus on aesthetics, some moments of ruptures, and cultural and artistic revolutions inside and outside Europe. The subject is then deepened in the third section, "Aesthetics and Sociospatial Inclusion in Postcolonial Cultures," through an analysis of how aesthetics—a philosophical discipline that emerged in the middle of the European Enlightenment to reflect on the beauty in works of art—can be interpreted and applied in the context of ex-colonies in artistic and literary works produced by the Indian diaspora in Southeast Asia (Myanmar) and Latin America (São Paulo, Brazil).

Overcoming (Post)Colonialism

The current context presents clear conditions of the complexity of the relationship between decoloniality, decolonization, and globalization, mainly considering concerns regarding social inclusion. As a consequence of the long construction process of modern knowledge and the sense of progress since the European Enlightenment, the most significant challenges for decolonization, according to postcolonial theorists, lie in the considerable fact that metropolis and colony are co-constitutive elements in a growing world capitalist order of development and civilization (Comaroff 2012, 6). Therefore, to think about the process of European colonization is to discuss the current context of the coexistence between former colonizers and the colonized, which resulted in the acculturation of native peoples. Moreover, the sense of "progress" became hegemonically linked to scientific, technological, and industrial development with its Cartesian bases and, economically, to the liberal market involved in the sociocultural transformation. The postcolonial theories that are critical to the consequences of the imperialist project include the search for a return to origins coexisting with contemporaneity as an alternative for a decoloniality of knowledge in Latin America (Mignolo and Walsh 2018), and also proposals for a decolonized future in Africa or by the African diaspora through Afropolitanism (Mbembe 2001). Through these two forms of worldviews, a historical panorama of Latin America and another one of Africa in meeting decoloniality and decolonization can be the basis for the meaning of postcolonial aesthetics. Here, "post-" is in the sense of a vision of the consequential effects of political and economic domination or control by knowledge in culture's formation, or considering the overcoming of subordination to this power. The difference between decoloniality and decolonization is that decoloniality does not succeed after colonialism and is understood as an alternative.

Decolonization implies overcoming colonialism, whereas decoloniality implies transcending coloniality, which concerns how development is operated by European modernity with objectives of world power domination. Thus, decoloniality, unlike decolonization, does not follow the cultural standardization of globalization as a reference but rather a hybridization. However, according to Arjun Appadurai (2021), it can be a reference for the rest of the world in seeking harmony with nature and different cultures. The author added that "this vision is seductive, and it is hard to disagree with their notions of harmony and conviviality, but it also rests on a reversal of the historical impact of capitalism and colonialism." Furthermore, the worldviews presented by Walter Mignolo and Catherine Walsh (2018) on Latin America and by Achille Mbembe (2001) on Africa and the African diaspora only reinforce the fact that nations, in the process of decolonization, are part of the global capitalist system in its development; thus, they are part of a worldwide system as such without having "truly fulfilled the liberatory promises of decolonization" (Appadurai 2021).

Nevertheless, these authors have presented a radical critique of European domination over the rest of the world, considering that materialist and Marxist analyses, with theories of dependency and world systems, have constructed both colonial and decolonial knowledge. In this sense, indigenous cosmologies present themselves as a form that opposes the values that underlie the capitalist system (Burocco 2021). Burocco argued that during the European civilizing process, African and indigenous cosmologies were presented as primitive; thus, they were considered obstacles to the project of modernity. Therefore, these cultures must coexist in a dialectical process today. Otherwise, to consider them as the colonizers did, conditioning them to a process of acculturation, would be to deny the contemporaneity of these native peoples. The author even exemplified the reaction of Ailton Krenak, a Brazilian indigenous leader, environmentalist, philosopher, poet, and writer of the Crenaque indigenous ethnic group, to the images of Indians captured by an English photographer in 1994; from his point of view, the images as a reference of the disappearing indigenous peoples were as if they were dinosaurs. Krenak claimed a vision present without translating Western epistemological and aesthetic parameters, which have no meaning in indigenous cosmologies.

Therefore, for Africa's decolonization, according to Mbembe, it is necessary to forget colonization. He believes that new technologies, added to new forms of language, art, and philosophy, are being experimented with to prepare this emerging Africa. However, from Appadurai's point of view (2021), Mbembe does not fully reconcile his optimism about Africa's political and technological future. The liberating future that could emerge from African decolonization lies precisely in not abandoning the complexities of modernity, including Africans' history as colonized. However, for Appadurai, Mbembe contradicts himself between the African past and future. In addition, Mignolo and Walsh, in their visions of a harmonious society, do not perceive the present violence and brutality nor think about the future consequences and promote a return to the precolonial past. Consequently, a critique of modern European thought becomes necessary as well as a contextualization of the centralizing power determining knowledge and, in this sense, the current aesthetic values through which all cultural and artistic production passes. Here, it is worth referencing the analysis of cultural spheres through anthropological and ethnographic research methods related to empirical postcolonial research in aesthetics. Both interpretative methods consider the symbolic action of observing behaviors in their differences. The analytical concern is the meaning, as in Clifford Geertz's *The Interpretation of Cultures* (1973). Geertz follows Max Weber's thought process by understanding the human being as an animal trapped in webs of meaning weaved by himself; thus, Geertz perceives cultures as webs and not as experimental research for determining results, but rather as interpretative ones in search of meaning. The structuralist methods consider nature and culture in their categories of cultural analysis developed by Lévi-Strauss. Furthermore, the very reference for postcolonial theories (Said 1978) stands out for poststructuralism as a method of research based on Michel Foucault, related to postcolonial reality to ground an epistemological conception (Kohn and Reddy 2017).

Simon During (1985, 1998, 2014), who significantly contributed to postcolonial studies, explained that postcolonialism emerged within literary studies during the 1980s as part of the general restructuring of the discipline. Consequently, literary criticism established effective literary judgments as a vehicle of cultural-political identities, resistance to ideology, and the impartial articulation of arguments in social structures. Critical influences on postcolonialism arose from the work of Edward Said (1978). Said followed Foucault to some extent in establishing an analysis of the relationship between power and knowledge, which provided new analyses that conditioned postcolonialism and its conditions of emergence. Recent postcolonial theories, in synthesis, present a critique of the Western production of knowledge. They consider a structured set of concepts, assumptions, and discursive practices used to produce, interpret, and evaluate knowledge about non-European peoples that transcend qualitative methodologies and often move toward an argumentative and activist critique. Consequently, to focus specifically on the importance of aesthetics in artistic production's meeting of the needs of the global South's indigenous and tribal peoples is to understand the importance of sociocultural inclusion in the current context. It is essential to consider the specific aesthetic categories applied to the artistic production of these peoples without any *parti pris* related to traditional European culture. The definition of "traditionally European" came about precisely with the definition of the "Oriental" as the other. Thus, Europeans came to have a conception of everything that is not Eastern through the process of "othering,"[1] defining themselves through the negation of a definition of what the other is. Therefore, the main challenge is to situate aesthetics in the postcolonial context since its origin is precisely in the European Enlightenment; consequently, it is linked to an elite concerned with categorizing and prescribing aesthetic judgments based on an ideal beauty.

Epistemological Rootedness and Aesthetics Toward Decolonization

Taking this brief notion of postcolonial theories in context with the emergence of the decentralization of knowledge, the meaning of the aesthetic theory faces some resistance by postcolonial theorists concerning the elitist values that still prevail in certain spheres of culture. However, some current aesthetic analyses focus on the universal appreciation of differences in the face of postcolonial cultural production. In this sense, postcolonial aesthetics follows the directives of this recent field of postcolonial studies. Furthermore, the qualitative cognitive aspects of postcolonial analyses enable transformations in aesthetic analysis based on transcultural aspects (Ashcroft 2015). The challenge in postcolonial aesthetics is related to the epistemological problems of aesthetics formation and development, and thus, to its object of analysis: Artistic beauty, that is, the sense of beauty or its reverse (horror) through the appreciation of the work of art. However, for this, we still require a previous definition of what art is. Here, the question is as follows: What are the criteria for establishing what is art or not? An inventory of achievements passes through our memory for classification, and in this sense, aesthetic judgment values based on the history of knowledge, whether philosophical, religious, or even common sense, are part of the Western Eurocentric formation and also, especially, the development of scientific knowledge. On the one hand, there is a Western historical legacy based on the European Enlightenment with a scientific and cultural background rooted in Cartesian reason. This thinking guides scientific, technological, industrial, and commercial development. Many thinkers have discussed the development of this knowledge in its usefulness or instrumentalization along this path. On the other hand, it is already known that otherwise generalized speculations without proof do not hold up. They may even cause controversy and much discussion, but they do not justify facts and truth.

Specifically in aesthetics, a first manifestation adverse to the Cartesian influence in the philosophical, political, cultural, ethical, and aesthetic fields predominated from the late 18th century to the early 19th century—namely Romanticism. Contrary to objectivism and the classical traditions focused on perfectionism—related to the beauty ideal—Romanticism was characterized by a worldview focused on the subjective aspects of the human being and freedom of thought. Both in Europe and its colonies during that period, movements arose against the condition of submission to monarchical rule. In Europe, the bourgeoisie conquered the public space and its domination through several revolutions. In the colonies, such as Brazil, the first manifestations emerged with decolonization; however, this highlighted a nationalist trend—a trend that represented Indianism in José de Alencar's paintings in the arts. Still, beyond this characteristic, the most important thing is that art, which had previously been exclusively focused on the nobility, started to value folklore—that is, popular art, an art linked to the people, of a particular ethnic population, to everyday life, transcending the aesthetic and prescriptive judgments of the elite.

Moreover, in opposition to the entire Western classical formation and even against the foundations of German Romanticism, Friedrich Nietzsche's work stands out. Nietzsche's reflections were focused on a critical analysis of the period in which he lived (1844–1900). In the last decades of the 19th century, sociocultural transformations were guided by the new technological, scientific, and industrial development processes through Cartesian reason. However, Nietzsche established the idea of finding a new meaning for humanity. As one of the precursors of nihilism, he established three stages of overcoming for humans to regain their autonomy and free themselves from the domination of the state and the church. Consequently, nihilism conditioning them as iconoclasts distanced them from any aesthetic contemplation by breaking with traditions and absolute truths regarding the effects of religious images and any power over humans. Nietzschean nihilism encompasses all human knowledge, from literature and art to the humanities and social sciences, passing through the spheres of ethics and morality.

The first of Nietzsche's stages is passive nihilism (*Passiver Nihilismus*), where existential meaning depends on orientation by beliefs to the most necessary moral and ethical attitudes of human relations. These include respect, equality, honesty, and tolerance. Then, however, it surpasses itself by realizing that the meaning of life can be something else by reaching the second stage, namely imperfect nihilism (*Unvollkommener Nihilismus*), which consists of the search for new meanings for life. This is when new ideologies are created, transmitting new guiding values, such as communism, socialism, Marxism, and liberalism. However, each of these ideologies pretends to be the absolute truth and orientation, causing divergence and instability for an ethical orientation. Doubt thus remains concerning the true meaning of life through belief, on the one hand, and ideology, on the other. Individuals consider that without these orientations, life would have no meaning. Everything would be allowed between good and evil, and finally, human would reach despair. This is where the third stage resides, namely active nihilism (*Aktiver Nihilismus*). Thus, achieving the ultimate consequences of nihilism means encountering "the nothing" (from the Latin *nihil*, nothing), or rather a new beginning, considering Western thought's principle, reestablishing a new meaning—a new worldview. This would be a challenge beyond human conditions, toward a higher dimension. For Nietzsche, reaching this "superhuman" position (Übermensch) consists of overcoming weaknesses, fears, beliefs, and eternal values. Therefore, this kind of human could reach the stage beyond this human condition, finding a new world and new values for human beings. In this sense, with a capacity beyond traditional values, they would find the foundations of new values to develop from this last stage of nihilism the means for their creations. So, they would reach the stage of the "will to power" (*Wille zur Macht*), but not in the sense of exercising power over others, but control over themselves—revaluing themselves.

However, what is the meaning of art concerning this reflection? For humanity to find an optimistic meaning in life, art—in its etymological sense—refers (in Greek to the term *téchne* and *ars* in Latin) to the sense of the techniques of mastering knowledge and the ability to perform intellectual activities in a particular way (see Chapter 4). Therefore, it is understood as artifice and invention; this art can create new worlds that are reliable. According to Nietzsche, we would be in a world of illusions, one of appearances (*Scheinwelt*)—an artificially invented world. In this sense, art is the solution. In a pre-Socratic period, Nietzsche developed his thought by seeking meaning in Greek antiquity before the Homeric poems. The Iliad is considered the oldest poem or the oldest record in Western literature. Thus, through a genealogy, Nietzsche sought to reestablish a whole new meaning for Western history by analyzing arts, especially tragedy, questioning the origin of Western values to understand the history of humankind's evolution through art. Western thought, since ancient philosophy, has always paid attention to art and technology as they are the shapers of culture. The impulse for humankind to produce, be it art or artifacts, is a necessity in the first instance. In the history of art and that of civilization itself, humankind in its evolution has always sought more than was provided in principle by nature—not only better survival resources but also better ideals for accepting the world offered to them.

In Germany, at the passage from the 19th to the 20th century, based on Nietzsche's work *Thus Spoke Zarathustra: A Book for All and None* ([1883–1892] 2005), the movement *Der Blaue Reiter* (The Blue Knight), founded by Kandinsky, opposed Cubism. Seeking individual values concerning reality, it emerged in the arts, in expressionism, with an emphasis on Franz Marc, Paul Klee, and the *Die Brücke* movement (German for "the bridge") in the paintings of Ernst Ludwig Kirchner and Karl Schmidt-Rottluff. The goal of this "bridge" was the passage from that moment to the future, bringing art closer to reality, abolishing the neo-romantic influences of that period in Germany, concerning the context of the Weimar Republic, post–First World War, from 1918 to the rise of Hitler in 1933. However, this moment only represented the beginning of a series of modernist manifestos and movements throughout the cultural sphere inside and outside of Europe. In particular, an approximation of art with industrial society was sought through manifestos, movements, and treatises. In Brazil, after Romanticism with the search for a national identity through the arts, *The Anthropophagic*

Manifesto by Oswald de Andrade (1928) emerged as a significant highlight and part of the modernist movement. It was considered a break with the values of art according to European and colonizing tradition, mainly referring to the models and standards copied by Brazilian society, that is, exercised aesthetic, cultural, and intellectual dominance over Brazilian culture.

The idea of anthropophagy was to use all of the circumstances of an independent Brazil, that is, one without European aesthetic influences, to lead Brazilian art production to its own identity, thus finding freedom of creation and a balance between national and international peculiarities. The references used by Andrade for the development of this manifesto comprised the leading European thinkers and artists of the time, such as Andre Breton and Francis Picabia, along with references from the Enlightenment, such as Jean-Jacques Rousseau, and also from materialism and psychoanalysis, such as Karl Marx and Sigmund Freud. This "mix of European ideas" ironically formed the basis for the movement that Andrade inaugurated with the objective of "rescuing the origins of Brazilian cultural formation," considering the primitive indigenous, African, and Latin cultures acculturated by European colonization. Alongside this mix of non-native but European ideas, he mistakenly used the term "anthropophagy" (cannibalism),[2] which characterizes a "Eurocentric" and colonizing posture of a distorted vision of native peoples as aggressive savages. Nevertheless, for renowned Brazilian anthropologists and sociologists (Miceli 2003; Freyre 2009), the Anthropophagic Movement does not represent a Brazilian specificity in a postcolonial aesthetic, even if they deny the European influence. To the so-called transition of Brazilian art, let us add an observation by Brazilian anthropologist Gilberto Freyre about those artists who contributed to the modernization. Freyre distinguished the modernization of Brazilian art by those who, according to him, "modernize, but are not modernists," indicating that the movement of the so-called "modernists" does not correspond to the modernization process of Brazilian art and culture. In a positive sense, Freyre referred to the music of Villa Lobos, who, as a follower of Johann Sebastian Bach, was revolutionary and independent with his Brazilian music. In poetry, Manuel Bandeira changed the forms of expression of the Portuguese language and thus had a notable influence on the literature of Jorge Amado and Graciliano Ramos. Finally, in painting, the works of Candido Portinari were valuable in the current Brazilian political and social scenario, as demonstrated in the previous chapters. In particular, results that portray distinct cultural mixtures of the country belong to Portinari's paintings.

In this sense, we can understand decolonization as a condition that does not replace but is an alternative to the creation of new meanings. It is considered the coexistence of different realities for decentralizing intellectual, cultural, and artistic power guided by modern European ideas that resist time, through new ideas and new dimensions of space, whether with an eternal recurrence in the Nietzschean sense going back to precolonial times or overcoming the effects of colonization. Thus, in the next section, significant theories on aesthetics are presented—with a focus on the Indian diaspora in Southeast Asia and Africa and the African diaspora in Latin America—as new ways to perceive, understand, and form a postcolonial society. This is in the sense of overcoming the construction of violence through the construction of otherness as difference as well as overcoming the negative consequences of globalization as a proposal for a consensual and reconciled future.

Aesthetics and Sociospatial Inclusion in Postcolonial Cultures

With the advent of the Third Industrial Revolution (or the technoscientific revolution) after the Second World War, technological development focused on the production process and scientific advancement. Cultural transformations were no longer limited to just a few countries but were global. They influenced not only production but also social relations, society's way of life, geographic space, and new aesthetic forms related to information and communication technologies. All of this development provided by the advances comprising all areas and activities of knowledge was related to globalization, including accelerating the dynamic relationship between time and distance, keeping people and places connected, transmitting information instantly, and seeking to overcome the challenges and obstacles that permeate the cultural, physical, and social differences concerning geographical locations and economic conditions. However, soon after the Second World War, when this revolution was taking shape, criticism of the rationalization of society was developed in the critical and aesthetic theories of the Frankfurt School (Adorno and Horkheimer 1947; Adorno 1970), as discussed in "Cultural Images: Real and Imagined Experiences" (Chapter 1), where the dialectical reasoning about the effects of the reason was discussed, revealing knowledge as the liberation of the human being from its condition of subordination to the system, and simultaneously keeping the human being in a logic of capitalist development. In this sense, the rationality in Adornian aesthetics (1970) deepened in the sense of art related to industrial culture—a term used to differentiate from the notion of popular culture. Introduced in co-authorship with Horkheimer (1947), the term focused on industrial production, the social and economic development of the liberal market, and globalization, such as the commodification of culture itself, including aesthetic values. These authors, also considering Herbert Marcuse and Walter Benjamin, represent the Frankfurt School, which developed critical social theories based on Kant, Hegel, Marx, and Weber, among others. Frankfurt School critical theory, in this sense, was fundamental in structuring the analysis of postcolonial aesthetics in *Native Intelligence: Aesthetics, Politics, and Postcolonial Literature* by Deepika Bahri (2003), enabling connections between the material aspects of artistic production and global capital flows.

Through the dialectics of Frankfurt School thought, cultural production was analyzed with a particular focus on the arts in its negative and positive aspects in assuming that capitalism appropriates and commodifies all features of existence, especially aesthetics. However, postcolonial theorists concerned with the meaning of aesthetics in the face of different cultures in the globalized world have analyzed Adorno's negative aesthetics, arguing that Adorno ignored the anticolonial aspects; that is, he maintained a Eurocentric view. In this direction, Dipesh Chakrabarty (2000) discussed how modern European thought is fundamental, albeit considering adequacy to the political context of non-Western countries. However, he added the need to attribute importance to aesthetics for education that concerns feelings, affections, and cultural habits appropriate to postcolonial ways of thinking. Nevertheless, the central question related to aesthetics in its development from modern reason to the postcolonial context continues as its primary object, artistic appreciation, and the sense of beauty. Here, not only among postcolonial authors but also since Kant (1788), the understanding that aesthetic judgment can be simultaneously subjective and universal implies a grounding in the *Critique of Practical Reason*. Moreover, the evolution of aesthetics as a discipline, that is, the way it has been developed and applied since Baumgarten, implies in each moment the historical, political, and cultural context of how the fundamental basis of reflection on beauty should be employed, undoubtedly considering the influence of Plato and Aristotle in their contradictory positions, complementary and essential to the entire development of aesthetic analysis. The exception was Hegel, who followed neither one nor the other, but presented a synthesis of these two bases for the construction of Western knowledge as a new worldview.

Worthwhile in this brief notion of the origin and development of aesthetics is to clarify that it is the abstract, philosophical reasoning proper to the idea of beauty that should be considered the foundation for knowledge. This knowledge can be acquired by any human being with access to education and training who can develop a critical and autonomous sense to think and understand the foundations of these theories. These theories have been interpreted and applied throughout their history in different contexts. Thus, it is a matter of how they are "adequate" and can thus be "partially" understood as adequate. However, to apply a fundamental understanding of these theories to the specific sociocultural context under analysis, these Western theories are related to the evolution of human knowledge, regardless of nationality, geography, or economic power. To further clarify what this appropriateness might look like, an insightful article by John Su (2011) offers an overview of how the aesthetic categories of the European Enlightenment are reclaimed and explored in postcolonial contexts, especially in Southeast Asia. In his analysis, the relevance of the aesthetic theories of Kant and Adorno is demonstrated as they are applied in Amitav Ghosh's literature. Furthermore, Su characterizes Chakrabarty's argument about the worldwide effects of European modernity on its development as radical. However, European knowledge is a nonlinear and coherent development through revolutions and ruptures. As such, Chakrabarty (2000) situates the problem of different ways of receiving this knowledge, considering that ontological assumptions based on modern European political and aesthetic thought can limit the understanding of postcolonial literature. From this statement, Ghosh's aesthetics is applied in a more "heterogeneous, fractured and fluid conception of modernity than

Adorno's and this is reflected in Ghosh's willingness to risk representing utopian communities" (Su 2011, 73). Ghosh's literature is an excellent fictional example of the limits of the language—whether verbal or visual—of the aesthetic experience and representation of beauty concerning Enlightenment thinking. The emphasis on Ghosh's work is significant for observing how modern European aesthetics are applied to postcolonial aesthetics related to the Indian diaspora. The connection between aesthetics and utopian thinking in Ghosh's work, especially in *The Glass Palace* (2001), presents a narrative of a modern Burmese family's history since 1989 in Myanmar. The main focus is on appropriation through aesthetic knowledge and education "to appropriate non-European others within a single dominant historical narrative that gives primacy to colonial empires and, more recently, to neocolonial forms of economic exploitation." However thank you specifically toward postcolonial aesthetics, Ghosh's work offers a significant utopian educational purpose regarding the notion of beauty and the possibility of communicating aesthetic experience while respecting cultural differences in the postcolonial context of a former British colony in Southeast Asia.

Next in this line of analysis, the application of the esthetic and critical theories of the Frankfurt School to Brazilian cultural production is considered. This analysis deals with a cultural production exclusively inserted in the Brazilian context due to industrial culture's dimensions in Latin America from the 1960s to the 1980s through appropriating and commodifying all spheres of human culture. It is based on the work *A literatura da cultura de massa: uma análise sociológica* (The literature of mass culture: a sociological analysis) by Waldenyr Caldas (1987), professor and researcher of Brazilian culture at the University of São Paulo. Caldas, in his research and analysis, identified values of Brazilian culture in the rhythm of the commodification project and leveled a criticism based on Adorno's considerations that valued the aesthetics of James Joyce and Samuel Beckett, with the idea of the high technical domain being possible for those with talent and competence; that is, it is accessible to those who know and can have industrial and technological development in favor of their ideologies, without necessarily surrendering to the system of superficial copies—namely kitsch. However, for Adorno's *Aesthetics* (1970), the reference of this high-quality and technical mastery was in the work of authors, artists, and musicians who sought, through modern art, a hermetic language subversive to the system, that only they—intellectuals—could understand as a form of subversion. Therefore, from the so-called mass literature, Caldas focused on the work of Adelaide Carraro. Unlike the literature consecrated by the academics who studied the most significant writers, such as Adorno considering the high performance of language in its hermeneutic meaning, Carraro does not fit into what is conventionally called great literature. Instead, Carraro's work belongs to a "literary stage" that, both in Brazil and Europe, is given the status of subliterature or *feuilleton*. The *feuilleton* appeared in France at the beginning of the 19th century and was imported to Brazil soon after, where it became a massive success in the second half of the century. It refers to a section, usually at the bottom of a newspaper's front page, where a writer would have a column in which they published a novel chapter by chapter, which could later be published in book form (today in new media, TV shows, and movie series). The theme of Caldas in question deals with everyday problems and the underworld of marginalized people in São Paulo's urban-industrial environment. In this direction, Caldas' analysis of Carraro's novels, a great writer of this (sub)literary style, demonstrated how she used a trendy language while bringing to the surface the perversities that occur in the underworld of the São Paulo metropolis. The best example of this is also the most recurrent in her work, namely daily conflicts among street dwellers as well as other situations that sufficiently characterize the lives of people who, for the most diverse reasons, are on the margins of the productive process. Carraro worked with a fictional narrative, one where fantasy and reality are always in dialogue throughout her prose. Positive and negative utopias are always present in her work, offering a dialectic of narrative. The aesthetic aspects are convergent and divergent in the urban scenario of the city. Rigorously, they are utopias that simultaneously present negative and positive experiences through the daily aesthetics of the characters. This socio-aesthetic experience that runs through Carraro's work in its widespread effects is focused on the subjective and existential aspects of daily suburban life.

Urban centers are undoubtedly the optimal setting for the interpretation of such situations precisely because of the impersonality, which David Riesman, Nathan Glazer, and Reuel Denney have already dealt with in American society ([1950] 2020) regarding modern conformity. Here, people in metropolises are "inner-directed" and "other-directed" to follow the goals, ideology, and tastes of the community where they live, especially in modern suburbs, seeking not consensus but rather approval, even of their neighbors, to ensure their existence and housing. Consequently, people abandon their "inner-direction," the essence of their own lives, thus defining the middle class. Therefore, Caldas understands Carraro's narratives as a negative utopia staged in a polarized system between the underworld and the elite. Through literature, Carraro expresses her subjectivities about a world of contradictions by mediating suburban and elite experiences to middle-class readers. The narrative's negative aspects of reality notions of the sublime and the ideal of beauty in the face of pleasure in its desired dimension are shown as positive aspects. For this analysis, Caldas is guided by Marcuse's ([1955] 1979) reflections on sublimation as an aesthetic experience, which means replacing what is desired through another experience in which this desire can be fulfilled or, better yet, materialized. For Marcuse, only art would allow individuals to express what, in principle, has been repressed. Therefore, through the aesthetic experience, they could find reconciliation with the world. The imaginary result through artistic representation manifests the negation of reality. Carraro's characters in subaltern conditions always seek forms of material achievements as a desire for social ascension. The lack of the beautiful is found in the sublime's sense of reason and experience. Pleasure in many everyday experiences can be found through Marcusian sublimation, that is, through material achievements as a fulfillment of desires. Carraro presents a diversity of narratives in which value judgments related to the aesthetics of everyday life constantly present themselves, thereby delineating racial and class differences.

In this sense, in *Falência das Elites* (Bankrupting the Elites), Carraro (1965) highlights a narrative through dichotomies, wherein the characters in her plots belong to the class of socially underprivileged, disqualified from the urban order, relegated, and forgotten by the state power (Vieira 2017). As the material aspects in Carraro's work are attributed to aesthetic judgments, moral judgments also stand out in narratives in which the ethical and aesthetic relationship of her plots define the reality of the underprivileged, associating them with the good through ingenuity, honesty, and kindness in the face of social challenges and obstacles concerning social differences. In several of her novels, aesthetic aspects are always present. One particularly striking aspect is colors being related to human desires that transport the narrative to an aesthetic linked to colonialism: "Yellow refers to the sun and gold, white to purity and European origin, far from creating a cultural mosaic, the colors create a symbolic system defining signifiers that refer to the condition of class, race and Family" (Vieira 2017).[3] In their dichotomies between colonized and colonizer, the narratives find the contrasts of the urban space in all its social differences, and fundamental human values and rights are only fiction or possibility for the elite. In this direction, Vieira (2017) added that

> the concept of family can only be extensive if there is the mark of whiteness [the color white is associated with the elite, the colonizer, the European]. The idealized family is nuclear, firmed on male authority, and has a color.

These associations as social and cultural characteristics are understood as values imposed by the coloniality of power.

However, these aesthetic judgments concern the perception of color through a native conception, which depends on self-definition, such as defining European peoples as white and the others as black, yellow, and red. Yet, this color-based classification is not precisely aesthetically constructed by artistic practice. Brazilian society, for example, adopts a racial classification based on the appearance of individuals. The official classification system of the Brazilian Institute of Geography and Statistics considers five categories for racial classification: Black, brown, white, yellow, and indigenous. However, in general, the population classifies using black and white. These classifications, both official and native, are much discussed as they are related to inequalities, discrimination, and racism. In this context, one perceives through the arts the representation of these problems related to the perception of color associated with socially assigned values, that is, social constructions. For this, the focus of aesthetics is not similar to that of sociology or anthropology in its analytical methods of approach; rather, it lies in perceiving in works of art the aesthetic value and beauty explicitly attributed to the work being contemplated, be it painting, photography, performance, or film, among many other artistic manifestations. Nevertheless, this subject is highly complex in the social sciences and includes research coordinated by UNESCO.

In addition, it is unanimously understood that the concept of race does not exist biologically; however, in the social sciences, this concept makes discrimination visible while simultaneously seeking to combat racism since it is a socially constructed classification incorporated into social and cultural relations. Throughout history, one can observe how these social constructions based on the classification of colors related to ethnic origins have been represented by visual narrative, mainly paintings, followed by photographs and then cinema. Therefore, an aesthetic reflection depends on appreciation and interpretation leading to feelings of pleasure, beauty, horror, or disgust, which are values strictly linked to culture and society for meanings throughout history, having gone through many different ways of considering what is beautiful and what is not. In this case, postcolonial aesthetics necessarily includes object politics in the analysis by considering that many artworks portray social injustices, representing aesthetic beauty through tragedy, as discussed earlier. Through these analyses of the literature of Carraro and Ghosh, it is most crucial to highlight the significance of urban space for the conception of public policies, and of resignifications aimed at cultural diversity and social inclusion in a postcolonial context. In this regard, understanding this process in the positive utopian sense of Mignolo and Walsh, as previously stated by Appadurai, seems to be the most appropriate direction. These novels are considered to influence the aesthetic effects, contributing to the mediation of the underworld with the elite or the colonized with the colonizer, thereby enabling a connection and mediation of values for the acceptance of this coexistence.

Differences are present in all situations of public space construction and seek the sense of an interculturality through the arts and aesthetics of everyday life, representing all its subtleties related to the decolonial movement. This is the possible meaning of postcolonial aesthetics when considering the overcoming of differences in alternative forms of cohabitation, with new perspectives concerning cultural diversity in urban spaces of the global South. Regarding the understanding of neoliberal urban space, aiming at decoloniality, Eurocentric studies have focused on peripheral capitalist cities' contradictions in search of a better experience of popular urban occupations, considering marginalized urban regions (Burnett 2021). To this end, decolonial theories and aesthetics address the core values and possibilities of sociospatial inclusion. Social responsibility is one of the primary characteristics of this creative activity in its functional, rational, and objective aspects, not to the detriment of the sensitive values of aesthetics but mainly based on its values. That is how desires and particularities are sublimated through the arts, resulting in the synthesis of reason and sensitivity to meet society's needs. It is precisely here that the potential of aesthetics is found in creating and recreating new habits and new representations through the arts to achieve the best possibilities for social inclusion through intercultural projects. Among the proposals for interculturality, Canevacci's research (2013, 2015, 2022) stands out for its analysis of how digital cultures transform the concepts of identities. Identities are analyzed according to an interdisciplinary ethnography and even an aesthetic reflection relating to the concept of ubiquity. The current digital context includes all identities as more flexible than the analog era. In Canevacci's method of research, "the ethnographic gaze has to be ubiquitous to decipher the conflicting codes (written, visual, musical, mixed, etc.) while practicing different dialogical narratives." Further, in abstract thinking for the city's configuration, Canevacci (2004, 2011) approximates an urban anthropology to the "wild thoughts" based on Claude Levi-Strauss' structuralism.

Therefore, the fundamental purpose of decolonial aesthetics and the whole decoloniality project is a liberation from the modern European sensibility tied to colonial values. The meaning is not focused on multiculturalism but rather on interculturalism, as one argument in the Decolonial Aesthetics Manifesto highlighted (Lockward and Mignolo 2012).[4] Moreover, a reflection on autonomy and the subalterns' power through the arts mediating coloniality and decoloniality can be attributed to thinking about the conditions set by Nietzschean Nihilism through overcoming the sense of the postcolonial context based on the illusion of domination by globalization. The main point defined in this discussion that answers the preliminary questions is the notion of beauty in its most profound qualities of the human condition. Further, it concerns the feeling of pleasure and, consequently, its inverse in the face of life's challenges and is related to the values proper to each culture, ethnicity, nationality, and class. These aesthetic values seek to form the new generations. Therefore, a rational aesthetic arises that is focused on the simultaneity of these two contradictory facts, namely the reproduction of inequalities coexisting with the growing disregard for racist behaviors, inequality, and the relationship with Eurocentrism, especially in the digital context of communication expanding the public space and thus resignifying the sociocultural dynamics. In this regard, the next issue, "Sociopolitical Facts, Art, and Technology" highlights the technological potential related to aesthetics and its rational principles. To this end, it presents an analysis of how sociopolitical factors are determined by rationality through art and media, seeks to understand the range of concepts related to technological developments and the role of human beings, and discusses the ability of humans to achieve positive results concerning their sociocultural circumstances.

Notes

1 John Harris and Vicky White: "Other/othering: Placing a person or a group outside and/or in opposition to what is considered to be the norm." *A Dictionary of Social Work and Social Care* (Oxford: Oxford University Press, 2013).

2 The Editors of Encyclopaedia Britannica:

> cannibalism, also called anthropophagy, eating of human flesh by humans. The term is derived from the Spanish name (Caríbales, or Caníbales) for the Carib, a West Indies tribe well known for its practice of cannibalism. A widespread custom going back into early human history, cannibalism has been found among peoples on most continents. Though many early accounts of cannibalism probably were exaggerated or in error, the practice prevailed until modern times in parts of West and Central Africa, Melanesia (especially Fiji), New Guinea, Australia, among

> the Maoris of New Zealand, in some of the islands of Polynesia, among tribes of Sumatra, and in various tribes of North and South America.
>
> Revised and updated by Adam Augustyn, https://www.britannica.com/topic/cannibalism-human-behaviour

3 Unless otherwise noted, all translations are my own.

4 Alanna Lockward and Walter Mignolo:

> Multiculturalism is managed by the State and some affiliated NGO's, whereas interculturality is enacted by the communities in the process of delinking from the imaginary of the State and of multiculturalism. Interculturality promotes the re-creation of identities that were either denied or acknowledged first but in the end were silenced by the discourse of modernity, postmodernity and now altermodernity. Interculturality is the celebration by border dwellers of being together in and beyond the border. Decolonial transmodern aesthetics is intercultural, inter-epistemic, inter-political, inter-aesthetical and inter-spiritual but always from perspectives of the global south and the former-Eastern Europe.
>
> "Decolonial Aesthetics (I). The Argument as Manifesto," *IDEA* 39 (2011), 89–97, reprint in *BE.BOP 2012. Black Europe Body Politics*, eds. Alanna Lockward and Walter Mignolo, Berlin: Kultursprünge im Ballhaus Naunynstraße, May 2012, 35–36

Manifesto. Signed by Alanna Lockward, Rolando Vásquez, Teresa María Díaz Nerio, Marina Gržinić, Michelle Eistrup, Tanja Ostojić, Dalída María Benfield, Raúl Moarquech Ferrera-Balanquet, Pedro Lasch, Nelson Maldonado Torres, Ovidiu Ţichindeleanu, Hong-An Truong, Guo-Juin Hong, Miguel Rojas-Sotelo and Walter Mignolo.

Bibliography

Adorno, Theodor. *Ästhetische Theorie* [Aesthetic Theory]. Frankfurt am Main: Suhrkamp, 1970.

Appadurai, Arjun. "Beyond Domination: The Future and Past of Decolonization." *The Nation*, 9 March 2021. https://www.thenation.com/article/world/achille-mbembe-walter-mignolo-catherine-walsh-decolonization/

Ashcroft, Bill. "Towards a Postcolonial Aesthetics." *Journal of Postcolonial Writing* 51, no. 4 (2015): 410–421. doi: 10.1080/17449855.2015.1023590

Bahri, Deepika. *Native Intelligence: Aesthetics, Politics, and Postcolonial Literature*. Minneapolis: University of Minnesota Press, 2003.

Burnett, Frederico L. "For the Decolonization of Urban Occupations: The New Factory Is the Neighborhood." Trans. by Brenda Veneranda Fernandes Silva. *V!RUS*, 23 (December 2021). http://www.nomads.usp.br/virus/virus23/?sec=4&item=9&lang=en

Burocco, Laura. "Reflections on the Future and Past of Decolonisation: Africa and Latin America." Third Text, Critical Perspectives On Contemporary Art and Culture. September 1, 2021. http://thirdtext.org/burocco-reflections

Caldas, Waldenyr. *A literatura da cultura de massa: uma análise sociológica* [The Literature of Mass Culture: A Sociological Analysis]. São Paulo: Lua Nova, 1987.

Canevacci, Massimo. *A cidade polifônica: ensaio sobre a antropologia da comunicação urbana* [The Polyphonic City: Essay on the Anthropology of Urban Communication]. 2004. Reprint, São Paulo: Studio Nobel, 2011.

Canevacci, Massimo. *The Line of Dust*. Canon Pyon: Sean Kingston Publishing, 2013.

Canevacci, Massimo. "Ubiquitimes: The Ethnographic Experiences of Digital Cultures and the Syncretic Mix of Space-Times." Paper presented at the seminary on Time, University of São Paulo, São Paulo, 2015. https://bit.ly/3txW8sN

Canevacci, Massimo. "Ubiquities: Aesthetic and Anthropology Glances on Undisciplined Identities." *Art Style, Art & Culture International Magazine* 9, no. 9 (March 2022): 11–25. doi: 10.5281/zenodo.6338582

Carraro, Adelaide. *Falência das Elites* [Bankruptcy of Elites]. São Paulo: L. Oren, 1965.

Chakrabarty, Dipesh. *Provincializing Europe: Postcolonial Thought and Historical Difference.* Princeton, NJ: Princeton UP, 2000.

Comaroff, Jean, and John L. Comaroff. *Introduction. Theory from the South: Or, How Euro-America Is Evolving Toward Africa.* New York: Paradigm Publishers, 2012.

During, Simon. "Postmodernism or Postcolonialism?" *Landfall* 39, no. 3 (Spring 1985): 366–381.

During, Simon. "Postcolonialism and Globalisation: A Dialectical Relation After All?" *Postcolonial Studies* 1, no. 1 (April 1998): 31–49.

During, Simon. "The Postcolonial Aesthetic." *PMLA/Publications of the Modern Language Association of America* 129, no. 3 (2014): 498–503. doi:10.1632/pmla.2014.129.3.498.

Freyre, Gilberto. *Modos de homem e modas de mulher* [Men's Manners and Women's Modes]. São Paulo: Global, 2009.

Geertz, Clifford. *The Interpretation of Cultures.* New York: Basic Books, 1973.

Ghosh, Amitav. *The Glass Palace.* New York: Random House, 2001.

Harris, John, and Vicky White. *A Dictionary of Social Work and Social Care.* Oxford: Oxford University Press, 2013.

Horkheimer, Max, and Theodor Adorno. *Dialektik der Aufklärung. Philosophische Fragmente.* Amsterdam: Querido, 1947. Reprint, Frankfurt am Main: Fischer, 1998.

Kant, Immanuel. *Kritik der praktischen Vernunft* [Critique of Practical Reason]. Riga: Johann Friedrich Hartknoch, 1788. Reprint, Stuttgart: Reclam, 2008.

Kohn, Margaret, and Kavita Reddy. "Colonialism." *The Stanford Encyclopedia of Philosophy* (Fall 2017). Edward N. Zalta, ed. https://plato.stanford.edu/archives/fall2017/entries/colonialism/

Lockward, Alanna, and Walter Mignolo. "Decolonial Aesthetics. The Argument as Manifesto." *IDEA 39* (2011): 89–97.

Marcuse, Herbert. *Eros and Civilisation. A Philosophical Inquiry into Freud.* Boston, MA: The Beacon Press, 1955.

Mbembe, Achille. *On the Postcolony.* Berkeley: University of California Press, 2001.

Miceli, Sergio. *Nacional estrangeiro: história social e cultural do modernismo artístico em São Paulo* [National Foreigner: Social and Cultural History of Artistic Modernism in São Paulo]. São Paulo: Companhia das Letras, 2003.

Mignolo, Walter, and Catherine Walsh. *On Decoloniality: Concepts, Analytics, Praxis.* Durhan e Londres: Duke University Press, 2018.

Nietzsche, Friedrich. *Also sprach Zarathustra: Ein Buch für Alle und Keinen* [Thus Spoke Zarathustra: A Book for All and None]. Leipzig: Ernst Schmeitzner, 1883–1892. Reprint, *Thus Spoke Zarathustra*, trans. Graham Parkes. Oxford: Oxford World's Classics, 2005.

Riesman, David, Nathan Glazer, and Reuel Denney. *The Lonely Crowd: A Study of the Changing American Character.* 1950. Reprint, New Haven, CT and London: Yale University Press, 2020.

Said, Edward W. *Orientalism.* 1978. New York: Vintage Books, 1979.

Su, John J. "Amitav Ghosh and the Aesthetic Turn in Postcolonial Studies." *Journal of Modern Literature* 34, no. 3 (Spring 2011): 65–86. https://doi.org/10.2979/jmodelite.34.3.65

Vieira, Adriana F. "Gênero, sexualidade e família nos romances da escritora Adelaide Carraro (1964–1985)." [Gender, Sexuality and Family in the Novels of Writer Adelaide Carraro]. Seminar paper, *História do Tempo Presente*, 3rd International Seminar, 2017.

7

SOCIOPOLITICAL FACTS, ART, AND TECHNOLOGY

Based on the discussions in the previous chapters, the meaning of the evolution of knowledge and technology in terms of its aesthetic and social implications is understood in Western and non-Western scenarios. Beyond the consequences of an intercultural space, the focus of this chapter is on rational aesthetics. Considering the science and technology that concern society implies accounting for how the disciplines that fall under this category affect their social environments and vice versa. A paradox inherent in this consideration stems from the conflict faced in the coming to consciousness of our existence as human beings. On the one hand, this involves acknowledging our mortality, an essential aspect of human life; thus, creating new models is a way to transcend our limitations. On the other hand, technology has revealed social and philosophical implications since the 19th century and the onset of the Industrial Revolution. Among these implications is the dehumanization that followed in its wake, which is a universally employed means of designating the effects of technology as it is shaped and structured to fit the new society that it has, in part, created. Dehumanization concerns the general impression of acceleration in technological progress and its attendant effects. These effects are often interpreted as problematic to life, because human beings are dependent on their functions. However, human beings have always depended on better social relationships and become increasingly reliant on them. The range of concepts that accompany technological development, such as the myth of progress and the role of human beings (i.e., dehumanization or the existence of the nonhuman), means that individuals increasingly demand identification and significance. Moreover, they imagine that their new creations imply enhanced communication, always relative to the continuous productivity in various sectors. In such ways, complex social concerns relate to the imaginations of different audiences, with forms of communication embodying the most appropriate relationship. Technology thus contributes to the ability to reproduce

DOI: 10.4324/9781003314943-8

ideas, improvements, and new developments. In the face of this growing scenario of new technologies and media, the following question is relevant: How do the visual forms of urban space interact with human reality? Some significant points considering sociopolitical facts, art, and technology are theoretically investigated in this chapter to seek clarification and an answer to this question (Weber 1922; Horkheimer and Adorno 1947; Wiener 1950; McLuhan 1964; Habermas 1968; Adorno 1970; Deleuze and Guattari 1972; Pinker 2018).

In the first section, titled "Aesthetic Education and Enlightenment: Beyond Technical Sets," the association between conceptions of the imagination and its variations is approached by analyzing inventions and artifices in cultural production and their relation to technological advancement. Therefore, through the universal art form of the moving image, which can be contextualized in the present day with cinematography, the aesthetics of film is analyzed, according to Schoolman's work (2020), as a medium that teaches us to overcome the construction of difference as otherness. A media analysis deduces possible knowledge of the transformation and characterization of the tendencies as a study method. The primary goal is to analyze the system of cultural production while extending the discussion of the industrial impact of automation from a significant and natural consequence to other fields, resulting in the perception of purposes established by human society regarding ethics, morality, reason, and practice. In the second section, titled "Art: Fact and Fiction," the development of knowledge of tendencies is discussed as a product of invention and artifice that followed in the wake of new cultural and technological practices and demands. In the third and final section, titled "Communication and Technical Interest," the focus is on analyzing transformations in societies and understanding how new technologies interact with everyday life. Attention is drawn to the relevance of the idea that, based on culture, humans can construct both formal and informal modes of communication to transmit knowledge and perspectives. This would lead the participants of a given community to develop values, beliefs, habits, and knowledge, or elements of culture expressed through human creativity and the participants' intellectual faculties.

Aesthetic Education and Enlightenment: Beyond Technical Sets

According to Norbert Wiener (1948), the founder of cybernetics, information is a term that designates "the content of what we exchange with the outside world to adjust to it," which makes us adjust without realizing it. The process of receiving and using information is the process of our adjustment to the contingencies of the environment. Communication and control form part of the essence of the inner lives of human beings and belong to the life of society. As is well known, human beings must adapt to society, shape themselves around an outside world, and become "real" when informed by creative skills. Technology offers communication possibilities that enable us to join a social environment and be renewed by it. Whether controllable or not, constant transformation is advanced by such technological development, which gives rise to new possibilities of communication experience in a sociocultural context.

Beyond theories of surplus value and the masses relating to science and technology in their sociopolitical implications, Karl Marx's concept of general intellect, found in a passage in the *Grundrisse* (Marx 1857–1861) titled the "Fragment on Machines," is relevant here. Slavoj Žižek's (2006) consideration of this concept and the notion of "multitude" is significant: "From the Marxian 're-appropriation of surplus-value,' the very formula of overcoming capitalist logic remains indebted to what it wants to abolish." However, the work of McLuhan (1964), which explored the impact of communication technologies at the human level, and that of Wiener (1948), who created the fundamental concepts of cybernetics and explored its main socio-philosophical implications, still play a role in discussions of the social impact of automation and its purpose in the intellectual and scientific construction of invention. Žižek's understanding and opinion of the subject differ from those of American thinkers, who mainly developed an analytical philosophy. Thus, as Žižek (2012) argued, "philosophy is inherently axiomatic, the consequent deployment of fundamental insight. Hence, all great dialogues in the history of philosophy were so many cases of misunderstanding." He described thoughts' influence as misunderstanding throughout the history of Western thought, culminating in a final contention that such influences are "without exception grounded in a productive misreading" as well as the following question: "Did not the entirety of analytic philosophy emerge from misreading the early Wittgenstein?" Recall from Descartes (1640) that consciousness in all its operations is conditioned by error, mistake, and illusion, as discussed in Chapter 2, "Urban Space and Collective Consciousness." Therefore, Cartesian doubt relates to consciousness's self-affirmation in the context of mistake and error; it continually seeks affirmations to account for everything not confirmed to be true. At the beginning of the 20th century and beyond, logical empiricism played a fundamental role in the work of the thinkers of the Cambridge and Oxford schools as well as in the Vienna Circle, leading to the development of cognitive science and an approach to mathematical philosophy.

Since ancient times, human beings have been determined to overcome the difficulty of treating what is unknown in the outside world through constructing an inner world and employing subjectivity, relative to coexistence with changes in the outside world. The relationships between human beings define their image of the world, which is constructed and continuously transformed. In this transition from the analog to the digital era, the constant transformation of social relationships is underway in the chain of existence. At present, a convergence in how human beings relate to machines is appearing. With the advent of human-operated machines and artificial intelligence in communication technology, innovative continuities of image flows are emphasized as differentiating procedures. However, these continuities converge at a common point, namely technological convergence. Techno-human nature is to consolidate updated images and transport them through history. Beginning with the first concept of human beings and technical evolution, scientific experiments have focused not only on the past or the future but also on present achievements. The relationship between human beings and techniques is greatly influenced by scientific research and its compatibilities and incompatibilities with technological convergence, which could entail a confluence of our senses. Thus, communication is beyond symbolic analysis; it is now an attempt to articulate new images as social facts and messages. The most relevant aspects of new technologies are their mediating functions (devices and inventions).

According to Jürgen Habermas in *Technology and Science as Ideology* (1968), his main work on this subject, since the earliest development of technical concepts by humans, scientists have gradually experimented. This has not been in a way that is limited to a definite past or an idealized future; rather, they have expanded the interrelationships of both through their results and conclusions. Aristotelian thought relates *téchne* with *praxis*; Habermas took up this relationship in society and analyzed it based on the interests of the human species, considering in particular two ways in which human beings can take action and achieve their aims. The first is through rational choice, which implies a systematized and strategic method. The other entails the interaction of social actors acting together, creating a common cause with consensual norms (*Geltungsansprüche*). Habermas thus developed his theory of reason, arguing for an understanding of technology and science as statements of beliefs. This is an essential aspect of Habermasian thought for the critical analysis of practical uses of technology and its social implications, such as the current trend of reducing life to sets of technical problems for experts to resolve. Habermas's reflections on and clarifications of Max Weber's (1922) rationality concern the choice of strategy for the use of technologies and the organization of systems according to the objectives of institutions that aim at world organization. We must especially consider who has mastery of this capitalist system, whether it is nature or society, and also who has control over technology and science. Politics are intrinsic to the individual's existence in society in the public space at the level of the citizen—man is a political animal, as Aristotle stated. Nevertheless, in the contemporary world, the citizen generally stays in touch with the public space through the media, whose association with relevant issues determines the public's responses and grants additional power to public opinion, as previously seen in Chapter 2, "Urban Space and Collective Consciousness." However, the results of innovation as a solution depend on political will. Habermas stated that the rationality of decision makers' discourse guides the collective conscience through their communicative actions. Here, the following questions arise: How can one be sure that these actions entail positive effects? What determining force, beyond the capital, is there for the mastery and control of these actions to guide them to achieve results?

Economic, technological, and industrial development enable new forms of communication that entail access to information and acquire a greater amplitude of knowledge. Processes of social transformation cause new necessities to appear within relations between individuals, fundamental rights and conditions of social inclusion, and new requirements (i.e., new rights). The diversity and complexity of human rights lie in difficulties based on morality and are, by extension, in common agreement with the rationalization of society. Therefore, it is widely acknowledged that the state and society connive in developing technologies and the sciences, which is advantageous for both parties. For this reason, rationality can be characterized as a science and a technique, a process of organizing elements and guiding progress, and also simple, productive power. Therefore, due to the negative consequences of capitalism, rationality was stigmatized and understood to be nothing more than a form of manipulation, as articulated in "Cultural Images: Real and Imagined Experiences" (Chapter 1). Consequently, the rationality illuminated by Habermas relates to ambiguity in Weber's thought: On the one hand, it embodies dissatisfaction in the form of critical reasoning about development's productive forces, while on the other hand it also justifies relations' production as the status of institutions that have adapted to fundamental objectives. Some meaningful analysis of the meaning of technological development can be found in the above-cited work of Habermas (1968), in which he analyzes the consequences of arbitrary scientific and technological development, where productive forces form new relations as the result of production.

Thus, the legitimate human interest in the technical control of nature functions as an ideology—a screen that masks the value-laden character of government decision-making in the service of the capitalist status quo. Unlike Herbert Marcuse, who regarded that interest as specific to capitalist society, Habermas affirmed the technical control of nature as a genuinely universal species interest; in Horkheimer and Adornos' *Dialectic of Enlightenment* (1947), the technical interests did not require social dominance (Bohman and Rehg 2017). The authors distinguished two meanings of the faculty of reason in its relation to absolute truth and knowledge. For both concerns, the reason is ambiguous and dialectical. On the one hand, it liberates human beings from subservience, bringing clarification (i.e., enlightenment). On the other hand, it conditions them to technocratic consciousness in the service of the development of capitalism and the economic interests of the dominant class. A way leading out of reason is planned, but its reversal is by no means inevitable. From technological development, Marcuse expected individual liberation, or the discharge of necessity, and sacrifices of energy and time. Still maintaining the strength of the *Dialectic of Enlightenment*, Adorno's *Aesthetic Theory* (1970), influenced by Kant, Hegel, and Marx, was a significant object of discussion in aesthetic reflection from 1970 until the beginning of the 21st century. Adorno, like Lukács, reinterpreted artistic creation throughout history, seeking an understanding of what relationships exist between modern art and the society that gave birth to it. Lukács (1972), for his part, rejected modern art, considering it the expression of the decadence of society. Adorno, however, defended it, arguing that it was able to resist the mastery of the culture of massification, differentiating itself from reproduced art, superficially perceived as kitsch in its simple and ephemeral forms in a mass culture that is dominated by technological development. The term "rational" in Adornian aesthetics was used to denounce the technical and scientific rationality of the social and economic development of the liberal market. This paradoxical interpretation of reasoning is decisive for understanding the heteronomy in Adorno's aesthetics. Rationality in the universe of the arts must be used to accomplish works of art. This art, in this period, is understood as modern art, which is, according to Adorno, to be understood as a work of art precisely because it is found to differentiate itself from other mass products.

However, the foundations of visual studies and aesthetics are fundamental to an understanding today. In this regard, Schoolman's work (2020) can be referenced, which adds to critical and aesthetic theories in political and communication sciences on current society's relationship. The object of analysis—the visual image—requires a study that progresses far beyond the cultural aspects of the relationship between art and culture. It is precisely through Schoolman's interpretation of Adorno's work, titled "First Bridge: Thinking with Adorno against Adorno," that one recognizes the need for critical theory. The said theory encompasses modern society in its analysis and includes the theoretical results that were considerably significant when ideological questions about capitalist society during the 20th century were discussed. Schoolman's interest lies in theorizing images of reconciliation as the pedagogical vehicle for aesthetic education in search of democratic enlightenment. Accordingly, this is the issue highlighted throughout his book *A Democratic Enlightenment: The Reconciliation Image, Aesthetic Education, Possible Politics* (2020). Based on Schiller's letters on aesthetic education and Adorno's aesthetic theory, Schoolman develops his theory of the reconciliation image through the cinematographic or moving image. This theory is built upon the challenge of "overcoming identity's construction of difference as otherness." In this regard, Schoolman develops and updates three essential concepts taken from Schiller's work for our context. The first concept is "the imitative (mimetic) aspect of reconciliation" through which identity contrasts with the "image of difference." It is the idea of the "democratization" process. The second concept is "aesthetic education." Finally, the third concept is "the image of reconciliation" through a universal art form, which according to Schoolman can be contextualized in the present day with film, with the moving image searching for democratic enlightenment.

Therefore, reflections on the concept of *mimesis* should be noted in Schoolman and Adorno's aesthetics as well as the influence of Kant (1724–1804) and Hegel (1770–1831), especially the latter in terms of the form and content of art. Hegel's art system was coherently grounded in the spirit of the time—*Zeitgeist*—through its translation of the idea into a conceptualization as material form. Concepts were universalized and, in a specific art, overcame subjectivity through being externalized and materialized. Hegel noted that art had life cycles (birth, development, and decline) that were outweighed by the particularities of each period of history. Symbolic art was overcome by classical art, which was ultimately surpassed by the romantic period. With its universal force of technical development and the evolution of the spirit seen in it in the form of the expression of ideas, poetry became increasingly dematerialized in an affirmation of the absolute. Therefore, considering Schoolman's analysis (2020), "Adorno's reconciliation image marks 'the birth of the new image,' as Deleuze brilliantly theorizes that event in Cinema I."

Thus, in a discussion between the present author and Schoolman (2021) on these issues regarding certain features of the reconciliation concept that he theorizes, the following question was formulated: From a Hegelian vantage point, would cinema among art forms capable of expressing values that affirm subjectivity's exteriorization as in its endless struggles with others for recognition when constituting its identity in the most fundamental ways be overcome? Indeed, Hegel thought of poetry as a form of artistic expression. However, in the present context, what would this actualization look like? Schoolman responded with "the temptation to decipher how the reconciliation image in film both illuminates and is illuminated by the experiences of consciousness and the history of reason in Hegel's *Phenomenology*." In this sense, Schoolman highlighted Hegel's importance in the history of philosophy on aesthetics as the image of reconciliation. However, without much room for a full analysis of this importance, he shortly clarified that "Hegel's reflections on absolute knowledge may be of assistance, where mind uniquely aware of its limitations achieves rational self-understanding," regarding Hegel's discussion on art, which he further explained as "a consciousness and self-recognition of mind's role and mind's forms in human history and, eventually, in the universe." Schoolman discussed Hegel's spiritualized idea[1] related to the formulation of the image of reconciliation in film, denying its mimetic character but emphasizing *mimesis* in the ideal representation of the idea's image of reconciliation. For him, "this is because both Whitman's and Adorno's concept of reconciliation as a time arrived and a time yet to come imbues its image with the capacity to represent instances of reconciliation."

Thus, it is possible to visualize reconciliation and witness its absence "to testify reconciliation as an ineffable achievement that can never be completely represented." For Schoolman, art would allow us to think of the reconciliation image in film, preserving the spiritual idea of differences; therefore, as an awareness of the absolute and as the sensible way, the absolute can appear as a democratic ideal. By contrast, Adorno's concept of imitation in modern art is seen as the disenchantment of the universe because it increases the distance between artistic appearances and social realities. An artwork in terms of Adornian aesthetics should not be concerned with realistic content, objectified critique of social facts with the specific subject, or the intention to express political ideologies while dissimulating intentional propaganda. For Adorno, a work of art should present, with its technical resources, the artist's mastery and talent, and the spectator should perceive that it is a work of art that represents our reality. Within Adorno's aesthetics, Schoolman finds the evidence necessary for an image of reconciliation that Adorno had confined at that time, with modern art, to an intellectualized minority, and which today can ultimately be found in cinematography as a universal form of art—namely the "reconciliation image in art" and the "reconciliation image in film. Art and film would be aesthetic analogues" (2020, 165). Therefore, discovering the image of reconciliation through film, Schoolman develops the concept of rational aesthetics on which he establishes his theory for positive enlightenment in "The Reconciliation Image in Film as Universal Art Form." This theory is exemplified with analyses of the films *The Help* (2011) and *Gentleman's Agreement* (1947), which portray violence related to racial and ethnic differences while simultaneously presenting an image of reconciliation that aesthetically educates the public on how the violence of building difference as otherness can be overcome. By explaining the reasons for democratic enlightenment, Schoolman introduces the political role to be played by education—teaching the ideal that reconciliation would realize. Then, arguing that such education has been offered historically by art, he theorizes "aesthetic education," one whose pedagogical media has become films. Within these observations, still in conversation with the author, Schoolman clarified (2021) that it is not a historical coincidence that the first proposal for aesthetic education appeared in 1795 with Friedrich Schiller's *Letters on the Aesthetic Education of Humanity*. At that moment, the most significant revolution was occurring on the sociopolitical scene with the first modern democracies, namely the French Revolution of 1789. Such aesthetic education had the moral task of teaching the ideal of reconciliation through the art environment. Schiller perceived the new arts and art forms that emerged in the early modern age to aesthetically represent the diversity of differences in the human and natural world in a realistic manner. Art was liberated from representing the idealized world and approached social and political reality. However, considering the aesthetic theories already discussed, the first theoretical formulation of art to perform the work of aesthetic education appeared in Schiller's work. Thus, in Schoolman's aesthetic education, it is understood that art today can also model the reconciliation of identity, difference, and diversity in the sensual world through the versatility of aesthetic forms. However, while the images address fact and fiction, the meaning would be for their value as art and images of reconciliation.

Art: Fact and Fiction

Human beings have employed different means to acquire knowledge ever since a consciousness of its existence concerning the world developed in the human species. Nevertheless, the primary goal was and has remained to achieve discoveries. The abilities to organize, classify, select, and analyze using reasoning as a method of inquiry and deduction lead to a sense of truthfulness that is not content with organized demonstration but demands precise analytical presentation. However, objective evidence of facts is absent in visual culture. This lack of representation and reproduction stems from the lack of rational analyses through aesthetic experience. Such rational analysis is possible only if a plan is implemented for reasoning out the passage of time and the location in the space of the aesthetic experience and interpretations governing the arts in history. Hence, understanding the arts and the sense of technique is to understand the history of human evolution. Western thinkers have focused on the concept of *téchne* (Chapter 4). This focus originated in the late 19th century with its revival of the ancient Greek τέχνη, meaning art and craft, and its adoption of the term "technology," which is related to scientific knowledge, incorporating the creation, theory, and praxis of technical skills. According to *The Oxford Dictionary*, the term "technology" itself has its origin in the Greek "*tekhnologia* or 'systematic treatment,' from *tekhnē*' art, craft' + -logia." Cultures have been centered around technology as a type of knowledge through techniques such as arts and crafts.

The tendency of humanity to produce art or artistic products must be understood as a necessity, evident in the history of art as well as that of civilization itself. A major goal of human beings is to overcome nature. Furthermore, this tendency is also evident in the applied sciences as well as advances in information technology and industrial development. Moreover, suppose that we measured the differences between previous technologies and current ones together with their relationships with their environing societies. We would see that all of our senses continually adapt to new technologies. If evolution and technical progress are measured, it is clear that individuals have always used the power of techniques to realize art in new forms. Moreover, according to Walter Benjamin (1939), the techniques and means of reproduction have varied throughout history, as articulated in "Democratization of Art and Culture" (Chapter 4), and that, in turn, led to a transformation in the social environment, such as the Industrial Revolution, the innovations and knowledge derived from it, and the increasing scientific discoveries and literature spread. Beginning with this moment, many popular works of fiction appeared and remained on the margins of progress. Hence, the wealth created by the Industrial Revolution also became a concrete reality in the form of gadgets, inventions, or stories for those who lived on the margins, who consequently came to feel included in society. Whether or not we believe that a message conveyed in a narrative appears real, its truth is not the truth of life; it remains a representation of interpretation or invention, techniques to demonstrate possibilities for world vision through music, painting, and photography or film. In the 19th century, much creativity was expressed through reproduction and increasingly influenced by commercial markets. In the 20th century, the emergence of the consumer society generated growth in social needs related to the demands of daily life due to automation and the resultant decreases in the workforce, which was replaced by machines and computers. Thus, new ideas appeared almost simultaneously in different parts of Europe and the United States, revealing sources of innovation and ideas for use by the culture industry. Industrial culture redistributes the burden of various technical functions, seeking above all a functionality of trade and the market. Society then submits to new achievements of development, adopting, for instance, new communication technologies, which have made radio, television, and cinema accessible through the Internet and social media. Consumer culture plays the role of a market amplifier for affluent societies.

Wiener's (1948) work *The Human Use of Human Beings: Cybernetics and Society*, which centers on the reception and utilization of forms of information, allows social participation to be understood in its micro- and macro dimensions. This process begins with communication from individual to individual. It then progresses from one community to another, from one society to another, and from a community or society to the individual and vice versa, with the constant formation of measures that can fit the dimensions idealized by the supposed outer world. Such measurements can be understood as rules, limits, standards, norms, and orders disposed and provided by the individual, who constantly compares and judges all values. Following this line of reasoning, the only motive for life in society is to participate in establishing laws, ethics, and morality. When analyzed, this participation process allows the individual to contribute to society, which can sum up individual factors. For example, if the information is considered a term that designates content (i.e., an idea), then when the argument is empty (i.e., without content), it has no value for measurement, canceling out participation in society. Individuals without information do not enjoy social participation, living their ideologies from the fundamental idea that all social processes regulating values and behaviors through exchanges of cultural values are social participation.

Each society is supported by the interdependence of subjective and objective values in social relationships, and societies receive and emit messages through gradual technological evolution. This leads to the conclusion, that technologies and their consequent environments succeed rapidly. Still, each environment prepares humanity for the following one (McLuhan 1964). Technologies take control as functions of training perception, bringing new technology's psychic and social consequences to awareness. Social transformation leads to the idea of dehumanization in the face of the development of industry and technology. The mastery of machines maintains us within the barriers of the human condition relative to the natural and the artificial, the real and the virtual, and among all things in which inventiveness stems from the intrinsic necessity to create in human beings. Such creation simulates, enables, and transforms society, leading it to a fundamental transformation of the human condition into a technological and programmed one. However, this condition relates to a humanity that maintains an identical itinerary of permanent goals, adapting itself to society amid revolution toward welfare and marking its epoch in the linearity of its time.

In the 20th century, science fiction began to focus on the social impact of science in Europe. In the United States, another school of science fiction originated that did not share the abstraction represented by European social science fiction. Under the influence of the American dream, screens and magazines were stamped with exotic objects that transformed American fiction into an invention of gadgets and things. These appeared, then, as literature for the masses, with fantastic and science fiction stories. Good science fiction depends on the ability to invent and predict human beings' future achievements. The significant influence of science fiction on humanity's future lies in the creation of everyday objects that come from ideas of an infinite world of discoveries and conquest at an indeterminate time. Science popularization through this means began in 1950 with Isaac Asimov, who exemplifies a science fiction author and an individual who could create ideas. Asimov devoted himself almost exclusively to science fiction literature, publicizing science's findings in his stories. In a talk he gave in 1968, he said that everyone writing science fiction by necessity was forced to make predictions. By the word prediction, he did not mean a statement that something would occur but rather the fact that it was even a possibility and could happen. This affirmation was due to the popularization of ideas that fueled the expectations of the masses and the cultural industry, which, attentive to the demand for new products, motivated both the market and science to invent them. This expectation of novelty in occurrences and products leads to an obvious conclusion regarding the possibility of cultural transformation through the emission and reception of new meanings with, of course, results expressed in the invention in the fields of arts and communication. Most recently, for instance, it has been possible to observe the promulgation of many scientific theories that lack scientific evidence but stimulate discussion between academics and agents of culture. One such example was developed in a published lecture by Julian Jaynes, based on his 1976 book *The Origins of Consciousness in the Breakdown of the Bicameral Mind*. More recently, the TV series *Westworld* appeared, which took inspiration from Jaynes's theory, exploring the clearing out of the human condition by means of artificial intelligence (AI). *Westworld* offers a portrait of human consciousness. Hence, in the sociocultural system, artifice and invention belong to science fiction and its products, derived from creators' ideas that mediate the functions of consumers and markets, encouraging progress and continually adding novelty.

Communication and Technical Interest

The creative and communicative interests of humankind allowed it to survive. Furthermore, since the onset of modernity, individuals have been characterized by know-how, which refers to a quality of an individual technician endowed with the expertise and proficiency that allow them to know how things are done. Underpinning the individual's technological evolution and know-how throughout history is the notion that humanity follows a path created by invention on a journey toward the fulfillment of dreams and achievements, which are artificially constructed through adaptation to technology. However, synthetic conditions can be established between human beings as users and technological services and products. They relate to the materialization of ideas as an overcoming characterized by the evolution of technologies to improve relations, albeit in the context of the mastery of social and ideological political systems. Technology can also be used to increase interest or ability. Thus, applications are employed to improve safety, protection, accompaniment, and freedom of movement or are misused in surveillance, invasion, disrespect, and obstruction of movement. All new technologies are tested in secret and used by states' intelligence services. Above all, they are considered essential techniques for human beings to overcome as a mere means of production and to become social beings with invention and artifice through community creation. However, people distinguished by ideals remain separated, employing such artifices as the diversity of culture and language.

In principle, the development of technology and new forms of communication is supported by economic and state power but is done for the strategic purpose of domination. The information elements are precious and kept as guarded secrets in the business world, where they are used to gain advantage in competition and world domination. States maintain national and international communication networks to promote control under political and ideological goals through their national policies. However, such elements are the inventions of those who are only interested in developing processes and products using new technologies without any political or ideological interest. Scientific culture cannot challenge the government by keeping anything that has the potential to be used by political power secret, because it is financially dependent and subordinate to the law. Relative to this attitude adopted by civilization toward communication technology, a technological revolution in creation and development is underway that goes beyond the understanding of the mass of people. What determines this production cannot predict the consequences in the sociocultural context.

Indeed, in *Anti-Oedipus* (Deleuze and Guattari 1972), it was found that "capitalism's command is utterly simple: connect deterritorialized flows of labor and capital and extract a surplus from that connection." Recently, we have seen highly technological but deterritorialized youth using new technologies within their social networks. On the one hand, their subjects represent a wide range of activities, from exhibiting financial interest in media action and making their pages dominant social media presences to, more significantly, creating fake news, which bears a solid antidemocratic trend. On the other hand, they are also searching for ideals of freedom through attention to democratic values. All of this is occurring in a tenuous and ephemeral penumbra of hope and desire for world justice, which is used to transcend limitations that may arise with the creation of new models. In this sense, rationality beyond any aim of communication and aesthetics, like determination, can create a solution by capitalizing on the belief in better conditions for human beings. As a parting thought, one can mention Aristotle. In the *Nicomachean Ethics*, he argued that it "is not just to explain the philosophy of the excellence for human beings but also to demonstrate specifically how human beings can lead lives of excellence as activity in accordance with practical and theoretical reason" (Archie and Archie 2003). It is always possible to understand evil as the consequence of misuse, that is, of the freedom of the will, permitting the use of reason for destructive purposes.

Jean-Jacques Rousseau (1712–1778) attributed this misuse of reason to the negativity of the values upheld by society, influences that can corrupt an individual. Rousseau, who is considered to have great importance among the thinkers of the French Enlightenment, also argued that the nature of the human will can be traced primarily to the natural passions, namely love of the self, love of one's neighbor, and desire for the preservation of the good and peace.[2]

All creatures display a drive toward self-preservation. Hence, for Rousseau, genuinely moral qualities can only be found in applying reason to human affairs and conduct. This reasoning "requires the mental faculty that is the source of genuinely moral motivation, namely conscience. Conscience impels us to the love of justice and morality in a quasi-aesthetic manner" (Bertram 2017). It may be for these reasons that certain artworks are consecrated by the passage of time, which has transformed them into masterpieces. Although the criteria for the aesthetic evaluation of artworks and the communication of their purpose are globally acknowledged, Western art has come to essentially be regarded as having an eternal presence expressed throughout its history due to its instantiation of the fundamental values of what is suitable for a human being. Thus, an audience without a specialization in a particular art considers the vital presence of masterpieces, which receive the benefit of intrinsic human nature with an understanding (*Verstand*) of the human condition in a Kantian sense. Along these lines, the recent publication of Steven Pinker's *Enlightenment Now: The Case for Reason, Science, Humanism, and Progress* (2018) reminds us of the human condition, which is fundamentally not rational. The thought of the Enlightenment was influenced by Bacon (1561–1626), Descartes (1596–1650), Spinoza (1632–1677), and Leibniz (1646–1716), among others, and Enlightenment thinkers did not ignore emotions. They only confirmed that we share an ability to be rational, seeking enlightenment in terms of considering the consequences for good. This thought forms the basis of Kant's affirmation of the categorical imperative, which formed part of the thought of the classical Enlightenment, in addition to his thesis that reason is above action because it grants the possibility of autonomy to the subject. For Kant, goodwill is a principle (fundamental law)—it is the obligation to think of the consequences of actions for the good of humanity and its freedom. However, Kant's freedom does not consist of doing what one desires but rather doing what is right. An individual and their reason may be the best master of their own life, which allows for respect for others. Individuals guide themselves to produce respect among others and to benefit everyone, without exception. Each person must act with fundamental respect for others by virtue of their autonomy.

Thus, Kant's categorical imperative, in "the third formulation of which in the *Groundwork of the Metaphysic of Morals* (the so-called formula of the kingdom of ends) recalls Rousseau's discussion of the general will in *The Social Contract*" (Bertram 2012), bears Rousseau's influence, an influence that can also be seen in the ideas of Hegel and Marx, as noted earlier. Moreover, contemporary epistemic conceptions of democracy refer to Rousseau's discussion of *The Social Contract*. However, a comparison of Rousseau's political philosophy with Weber's sociology provokes the problem of legitimacy at an international level.[3] Rousseau examined legitimacy concerning the purpose of government and conduct. However, if legitimacy is to be related to the inherence of power in traditional legitimate domination, as formulated by Weber, the rationalization of social structures to preserve dominion by means of references that meet the fundamental interests of the state and the collectivity is to be found in the creation of a global culture and its values. Moreover, the values of art and literature are not unrelated. On the contrary, they form expressions of natural human feelings in their most subjective aspects, shaping the communication process with society, which today passes through the cultural industry. It is of great value to emphasize that politics is contained within this. Generally, rationality transcends aesthetics and communication concentrates on positive aspects of development. Reason's primary objectives express themselves as science and technology as progress, in the sense of productive powers that politically and economically serve global interests. Hence, this chapter has examined the complexity of reason's role in social life, especially its questionable instantiation in the use of science and technology. However, despite all of the discussions of instrumental rationality in the entire course of theories, which have thoughtfully engaged with the issue in its entirety, the understanding of reason remains a great enigma. However, humankind has always lived throughout its history despite lacking an absolute mastery of reason. It is essential to affirm that most inventions, at their origin, do not play a destructive role in society but rather are created to meet the needs and conditions of human survival. Accordingly, in Chapter 8, "Technicization and Aestheticization of Social and Urban Structures," the existing institutional relationships—the producers and managers of new technologies, media, and artistic and cultural practices—are presented as the forms of organization necessary for a technically rationalized society.

Notes

1 Morton Schoolman:

> I cannot now adequately defend my argument in Hegel's terms. I would be inclined to argue that the concept of reconciliation in *A Democratic Enlightenment* qualifies as a Hegelian 'Idea' that appears in a sensible form—the artwork—as an 'Ideal.' For example, the Idea of difference as democracy's highest spiritual value appears as an Ideal in the reconciliation image. As the sensible form of the Idea of reconciliation, the Ideal reconciliation image in film represents the emancipation of difference from its construction as the Other through a mimetic transformation of identity, becoming different in the image of difference. As the Ideal expresses an Idea of reconciliation, the reconciliation image in film represents the transformation—'subjectivity's overcoming its exteriorization'—of identities belonging to individuals, groups, communities, nations, genders, races, and sexualities, among other differences.
>
> "A Discussion on a Democratic Enlightenment: The Reconciliation Image, Aesthetic Education, Possible Politics." Interview by Christiane Wagner, *Art Style, Art & Culture International Magazine*, v. 8, no. 8.1, part 1 (September 2021): 37–55. DOI:10.5281/zenodo.5448260

2 Christopher Bertram:

> *Amour de soi, amour propre* and *pitié* are not the full complement of passions in Rousseau's thinking. Once people have achieved consciousness of themselves as social beings, morality also becomes possible, and this relies on the further faculty

of conscience. The fullest accounts of Rousseau's conception of morality are found in the *Lettres Morales* and in sections of the *Confession of Faith of the Savoyard Vicar*, a part of *Emile*. In the most primitive forms of human existence, before the emergence of *amour propre*, *pitié* balances or restrains self-interest.

"Jean Jacques Rousseau," *The Stanford Encyclopedia of Philosophy* (Winter 2020). Edward N. Zalta, ed. https://plato.stanford.edu/archives/win2020/entries/rousseau/

3 Merquior, J.G.:

Rousseau and Weber are as different in spirit as they are distant in time. … Nevertheless, the very contrast between our two theorists yields a fairly illuminating perspective on the concept of legitimacy. Indeed, each of them might very well be taken as the supreme representative, the archetype, so to speak, of one of two basic ways of looking at the phenomenon of legitimacy: that which views it in terms of belief, and that which sees it in terms of power. While Jean-Jacques Rousseau was the main founder of what might call the 'power theory of legitimacy,' Max Weber remains the *locus classicus* of the 'belief theory of legitimacy.'

Rousseau and Weber: Two Studies in the Theory of Legitimacy (London and New York: Routledge, 1994)

Bibliography

Adorno, Theodor. *Ästhetische Theorie* [Aesthetic Theory]. Frankfurt am Main: Suhrkamp, 1970.

Archie, Lee, and John G. Archie. "What Is the Life of Excellence? By Aristotle." In *Introduction to Ethical Studies*. An Open Source Reader, 2003. https://philosophy.lander.edu/ethics/ethicsbook/book1.html

Aristotle. *Nicomachean Ethics*. Trans. by W. D. Ross. Oxford: Oxford University Press, 1925.

Benjamin, Walter. "Das Kunstwerk im Zeitalter seiner technischen Reproduzierbarkeit" [The Work of Art in the Age of Mechanical Reproduction]. 1939. Reprint, *Medienästhetische Schriften*. Frankfurt am Main: Suhrkamp, 2002.

Bertram, Christopher. *Rousseau and the Social Contract*. London and New York: Routledge, 2004.

Bertram, Christopher. "Rousseau's Legacy in Two Conceptions of the General Will: Democratic and Transcendent." *Review of Politics* 74 (2012): 403–420.

Bertram, Christopher. "Jean Jacques Rousseau." *The Stanford Encyclopedia of Philosophy* (Winter 2017). Edward N. Zalta, ed. https://plato.stanford.edu/archives/win2020/entries/rousseau/

Bohman, James, and William Rehg, "Jürgen Habermas." *The Stanford Encyclopedia of Philosophy* (Fall 2017). Edward N. Zalta, ed. https://plato.stanford.edu/archives/fall2017/entries/habermas/

Deleuze, Gilles, and Félix Guattari. *Capitalisme et schizophrénie. L'anti-Œdipe* [Anti-Oedipus: Capitalism and Schizophrenia]. Paris: Minuit, 1972.

Habermas, Jürgen. *Technik und Wissenschaft als Ideologie* [Technology and Science as Ideology]. Frankfurt am Main: Suhrkamp, 1968.

Habermas, Jürgen. *Theory of Communicative Action*. Trans. by Thomas McCarthy. Boston, MA: Beacon Press, 1984. Originally published as *Theorie des kommunikativen Handelns* (Frankfurt am Main: Suhrkamp, 1981).

Habermas, Jürgen. *The Structural Transformation of the Public Sphere: An Inquiry into a Category of Bourgeois Society*. Trans. by Thomas Burger. Cambridge, MA: MIT Press,

1991. Originally published as *Strukturwandel der Öffentlichkeit. Untersuchungen zu einer Kategorie der bürgerlichen Gesellschaft* (München: Luchterhand 1962. Reprint, Frankfurt am Main: Suhrkamp, 1990).

Hegel, Georg Wilhelm Friedrich. *The Phenomenology of Spirit.* Trans. by Terry Pinkard (Cambridge Hegel Translations). Cambridge: University Press, 2018. Originally published as *Die Phänomenologie des Geistes* (Bamberg and Würzburg: Joseph Anton Goebhardt, 1807).

Hegel, Georg Wilhelm Friedrich. *Lectures on the Philosophy of Art.* The Hotho Transcript of the 1823 Berlin Lectures. Trans. by R.F. Brown, Oxford: Clarendon Press, 2014. Originally published as *Vorlesung über die Philosophie der Kunst.* Berlin 1823. Rewritten by Heinrich Gustav Hotho (Hamburg: Felix Meiner, 1998).

Horkheimer, Max, and Theodor Adorno. *Dialektik der Aufklärung* [Dialectic of Enlightenment]. Amsterdam: Querido, 1947. Reprint, Frankfurt am Main: Fischer, 1998.

Kant, Immanuel. *Kritik der reinen Vernunft* [Critique of Pure Reason]. Riga: Johann Friedrich Hartknoch, 1781, 1787, 1790. Reprint, Stuttgart: Reclam Verlag, 2006.

Kant, Immanuel. *Kritik der praktischen Vernunft* [Critique of Practical Reason]. Riga: Johann Friedrich Hartknoch, 1788. Reprint, Stuttgart: Reclam, 2008.

Lukács, Georg. Ästhetik [Aesthetics]. Berlin and Neuwied: Hermann Luchterhand, 1972.

Marcuse, Herbert. *Eros and Civilisation. A Philosophical Inquiry into Freud.* Boston, MA: The Beacon Press, 1955.

Marx, Karl. *Grundrisse.* First published 1857–1861. London: Penguin, 1973.

McLuhan, Marshall. *Understanding Media: The Extensions of Man.* New York: McGraw-Hill, 1964.

Merquior, J.G. *Rousseau and Weber: Two Studies in the Theory of Legitimacy.* London and New York: Routledge, 1994.

Pinker, Steven. *Enlightenment Now: The Case for Reason, Science, Humanism, and Progress.* London: Penguin, 2018.

Schoolman, Morton. *A Democratic Enlightenment: The Reconciliation Image, Aesthetic Education, Possible Politics.* New York: Duke University Press, 2020.

Schoolman, Morton. "A Discussion on a Democratic Enlightenment: The Reconciliation Image, Aesthetic Education, Possible Politics." Interview by Christiane Wagner, *Art Style, Art & Culture International Magazine* v. 8, no. 8.1, part 1 (September 2021): 37–55. doi:10.5281/zenodo.5448260

Wagner, Christiane. "Rationality: Beyond Aesthetics and Communication." *Conference Proceeding, 1st International e-Conferenceon Studies in Humanities and Social Sciences. Center for Open Access in Science – COAS* (June 2018): 1–12. https://doi.org/10.32591/coas.e-conf.01.01001w

Weber, Max. *Wirtschaft und Gesellschaft: Grundriss der verstehenden Soziologie* [Economy and Society: Outline of Understanding Sociology]. Berlin: Peter Kietzmann, 1922. https://www.textlog.de/weber_wirtschaft.html

Westerman, Richard. "The Reification of Consciousness: Husserl's Phenomenology in Lukács's Identical Subject-Object." *New German Critique* 111, v. 37, no. 3 (2010). https://doi.org/10.1215/0094033X-2010-016

Wiener, Norbert. *The Human Use of Human Beings: Cybernetics and Society.* Boston, MA: Houghton Mifflin, 1950.

Žižek, Slavoj. Multitude, Surplus, and Envy, Rethinking Marxism. London and New York: Routledge, 2006.

Žižek, Slavoj. Organs without Bodies. On Deleuze and Consequences. London and New York: Routledge, 2012.

8

TECHNICIZATION AND AESTHETICIZATION OF SOCIAL AND URBAN STRUCTURES

New techniques are continually being reported, in which one can always see similarities with past achievements. New technological devices and features implement material aspects of physical reality. Specifically, regarding the term "technicization,"[1] the analysis in this chapter is directed at the modern condition of humanity, in which technical standards of efficiency shape all aspects of life due to the development and technical improvement in all areas of human achievement. Furthermore, in the analysis of the process of aestheticization,[2] it should be noted that the term implies the context in which it is being used. It is due to the very conception of aesthetics in its historical development, which has undergone sociocultural transformations, that new values have attributed meanings. In this discussion, the term "aestheticization" is associated with the contemporary context, which involves the general and superficial aspects of aesthetic experience. Wolfgang Welsch (1996) characterized this phenomenon of aesthetics of everyday life as an extension of kitsch, describing the superficial values associated with pleasure and desire in cultural activities, which are mainly related to economic strategies. However, significant in this analysis is that the superficial experience in this aestheticization process is now reaching more profound layers of contemporary society. The bases are precisely in the productive processes and new technologies; therefore, "aesthetic processes don't only shroud already completed, given substances, but even determine their structure, affect the cover and even the core," according to Welsch (1996, 4). He also stated that "aestheticization is by no means merely a thing of *beaux espirits*, or the postmodern muse of amusement" (4), considering that it is a matter of "superficial economic strategies, but very much from fundamental technological changes, from the hard facts of the productions process" (4). On the one hand, this chapter refers to the material world that has been perfected through a technicization process related to science and technology development. On the other hand, the discussion presents

DOI: 10.4324/9781003314943-9

an economic system responsible for managing and investing in this technoscientific evolution. Therefore, this analysis considers that capitalism and its forms of commodification induce "superficial pleasure" and transform everything into merchandise, including the whole social and public spheres. Consequently, the world presents itself with two material faces: One is growing increasingly beautiful as it becomes more prosperous, while the other is growing uglier as it becomes poorer, as Figures 8.1 and 8.2 show.

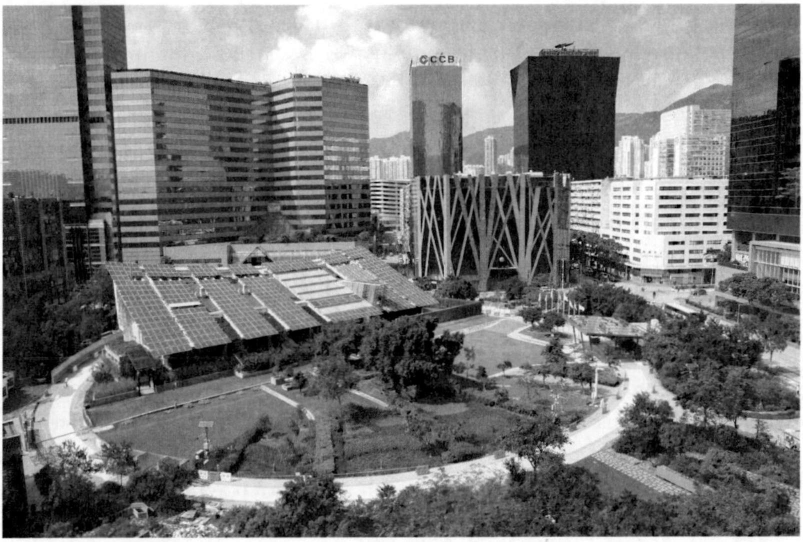

FIGURE 8.1 Image of a prosperous city.

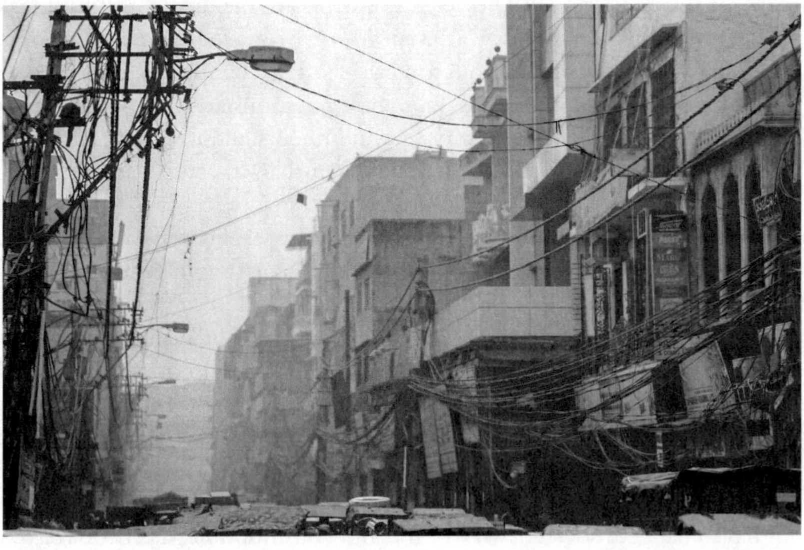

FIGURE 8.2 Image of a low-income city.

Pleasurable and superficial appearances involve daily urban experiences. In this sense, this discussion approaches Welsch's consideration that "aestheticization is at its most obvious in urban areas, where almost everything has been subjected to a face-lift over the last few years" (2). He described all of the structures in urban centers and public spaces, including the peripheries and the countryside cities. Welsch added that "even ecology has, largely, become another means of enhancement. If advanced Western societies were able to do completely as they wish, they would transform the urban, industrial and natural environment *in toto* into a hyperaesthetic scenario." Drawing on this assumption of Welsch regarding "the Western project," this discussion addresses the influence of technicization and aestheticization as image cultures[3] considering the challenges of the Fourth (Industrial) Revolution—the digital age mastering all reality. Despite this emphasis on a technological future that involves many imaginations about the human condition, human behavior remains the same at its essence. Moreover, individuals continue to struggle for basic needs in terms of fundamental rights due to different economic conditions between classes and within technologically developing and developed societies. In addition, humanity remains politically and socially oriented toward investments in new technologies aimed at technical and scientific evolution, striving to overcome Cartesian influence. Scientific knowledge has developed regarding general and particular cases through representations and applications as the complexity of the phenomena is solved through the integration of new theories. Thus, the formation of scientific knowledge is understood as a continuous process through conceptual syntheses to understand reality. However, given the challenges of science in the face of discoveries about our existence, it seems that the achievements of modern Western science are due mainly to the continuous search for knowledge and questioning of past human mistakes for future successes. These achievements are the most significant legacy of the Cartesian method. Thus, Cartesian reason protected humanity from natural realities and insecurities through methods and techniques of a mechanism that is no longer applied. Here, the following questions arise: How is the world moving forward? How can humanity live and understand its reality better? The technicization and aestheticization of social and urban structures are supposed to offer a better life, but as a constant promise that dissuades reality. To this end, this analysis seeks clarification through a retrospection of technological development and considerations of aesthetic aspects (Welsch 1996) that involve urban planning and visual configurations (Giedion 1978; Picon 2005–2006).

Development implies changes—that is, revolutions. These changes mean new economic and technological development stages that consolidate the capitalist system, which also implies new ways of organizing society and, consequently, public space. These changes transform all social structures, revolutionizing behavior and forms of organization and production. As a result, advances have been made in various fields of knowledge, productivity has increased, work relations have changed, society's way of life and consumption patterns have transformed, and the relationship between humans and nature has especially changed.

Since the 18th century in England, the Industrial Revolution has developed, presenting phases of significant changes related to technological and socioeconomic evolution, according to productive advances in the scientific field and several other areas of the economic and industrial sectors. Three phases have already been consolidated.

The first phase, namely technological advancement for industry, was focused on textile production with the technological development and creation of the locomotive, railroads, and telegraph throughout the United Kingdom and worldwide. Critical social changes also occurred due to industry replacing hand production. The second phase refers to the period from the second half of the 19th century to the middle of the 20th century, ending during the Second World War. It was characterized by significant technological advances and improvements to existing technologies, which increased industries' productivity and profits. These industrial activities are associated with petroleum as a source of energy and combustion engines, and then electricity began to be used as power for motors. The substitution of iron for steel was another significant achievement. The third phase, known as the Technoscientific Revolution, began after the Second World War, providing biotechnology, robotics, and advances in areas such as genetics, telecommunications, electronics, and transportation. This phase also transformed social relations and the relationship between time and space with globalization, thereby consolidating financial capitalism. Therefore, financial capitalism undoubtedly contributed to the continuous investment in technology and science, thereby promoting evolutions toward the digital era and configuring the fourth revolution.

The Fourth Industrial Revolution, also called Industry 4.0, encompasses a system of advanced and connected technologies comprising robotics, artificial intelligence (AI), machine-to-machine connections, big data, the Internet of Things (IoT), cyber security, 3D modeling, and cloud computing, which are changing the methods of production and the patterns of urban life. In urban projects, simulation systems with computers and techniques are used to generate digital models that enable complex interactions and applications in virtual and augmented reality, thus imitating real-world processes. Moreover, the entire socioeconomic structure has been fully digitalized in its productive, educational, and cultural aspects. This rationalization of the system in its course of industrial development, considering the changes in urban space, implies the technicization and aestheticization of social and urban structures—that is, how society is technically and aesthetically reconfigured in the face of Industry 4.0. In this evolutionary sense, this discussion aims to clarify the articulations focused on inducing the adoption of digital technologies and thus to analyze the advantages and disadvantages of this use by today's society and that of the near-future. In this direction, the first section, titled "The City: Technical, Functional, and Aesthetic Considerations," delimits the emergence of social projects for achieving essential goals toward sustainable development, aiming for solutions to inequalities and housing problems. The second section, titled "Smart Cities and Visual Configurations," discusses

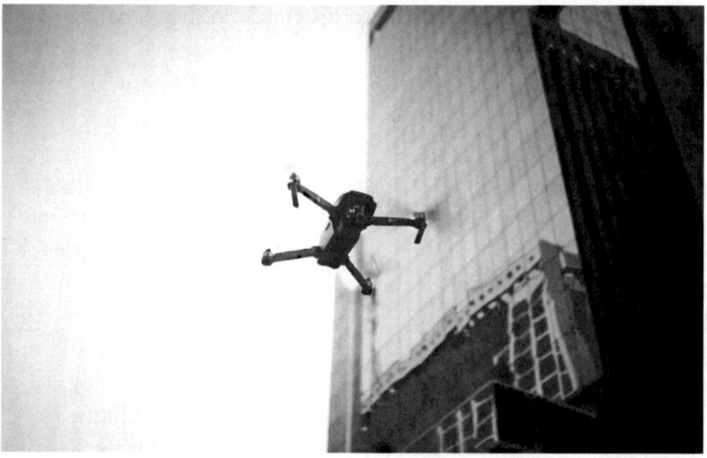

FIGURE 8.3 Apparent drone on surveillance mission.

the future of ideal cities under digital technologies' tutelage and certain urban design utopias. The final section, titled "The Digital Age and the (Dis)appropriation of Urban Space," presents examples and discussions regarding the forms of technicization and aestheticization of the diverse values of cultural reality. Furthermore, it highlights the challenges of cyber security in the current global context as well as the challenges that individuals face in a city under digital surveillance, as Figure 8.3 shows.

The City: Technical, Functional, and Aesthetic Considerations

Urban life, in its complexity, is the essential part of many cities, where people live together, building and rebuilding their histories as well as their moral, ethical, economic, cultural, social, and political values. Many cities aesthetically represent the space and time of this transforming urban life. When thinking about time and space in the context of practical and aesthetic achievements in the city, what matters is the perception of urban space and everyday life experiences related to the past and the future. Thus, the principal goal is facts linked to the growth of cities and projects that can meet social needs, solving problems such as housing, transportation, overpopulation, signalization, and communication. However, it must be considered that the first solutions appear as a result of the emergence of problems at various levels. Urban centers are still attempting to solve these problems due to the outdated industrial technologies and economic development from the second half of the 20th century until the present day. In addition, as already discussed, other problems concern the economic growth worldwide related to the differences between Western and non-Western nations. These economic differences are the main empirical factor in the relationship between time and space in urban transformation as well as the challenges for an improved quality of life in many cities, including the aesthetic, ethical, ethnic, moral, and political experiences. However, considering that city growth has been a phenomenon since the emergence of civilization, it is only after the Industrial Revolution, with the development of industrial centers, that many cities started to experience growth and urban development centered on commercial life. Time, therefore, is more related to the ability to direct the development of techniques to solve projects that can be managed according to the growth of urban centers. It is the relationship of space with the city's transformation both functionally and aesthetically.

According to Sigfried Giedion (1978), the conception of space does not occur independently and autonomously but rather through the organization of forms in space. In addition, cultures are the presence, influences, and heritage of humanity. This heritage can be universal or restricted to a small community. This dynamic, through which the individual acquires a presence as a citizen, operates under the influence of achievements, transmitting tradition, education, knowledge, history, art, and customs. However, the diversity of cultures, whether between nations or different nationalities concentrated in the same country, has an intellectual, spiritual, and aesthetic development in its process of acculturation. This is the case for the city of São Paulo, Brazil, which represents the image of a global metropolis, in addition to being seen as a booming economic powerhouse. However, São Paulo still faces social inequality in many aspects, which mainly affects the education of the economically most modest classes of the population. It is worth remembering that the presence and influence of a Brazilian and global metropolis, such as São Paulo, cannot be compared with those of European metropolises. First, the history of São Paulo began when it was a European colony of Portugal and was exploited for no less than three centuries. Second, different to a hegemonic metropolis, São Paulo did not profit from the two great wars of the 20th century through economic growth. Therefore, since the Second World War, the development of São Paulo and the whole capitalist world followed a Fordist model of industrial growth with a focus on technology, innovation, and consumption.

Today, issues of urban life as consequences of industrial development focus on the problems of overpopulation, lifestyle, transportation, and housing. Besides public policies and social projects, urban designers and architects project solutions for these problems. Moreover, the main factor in the evolution of science and technology is industrial modernization and design. Therefore, the relationship between art, design, architecture, and social reality is complex in terms of technological and sociocultural development. Thus, a visual analysis of the social context as aesthetic experiences of the social actors has become essential. Under these conditions, media images produced as part of the urban space are part of the complex system of visual culture, which implies current ethical, social, and political values. These media images are aimed at an innovative, creative production that interacts with diversification, thereby restructuring the fragmented values of culture's inherent differences and similarities, integrating sustainability across the value chain to ensure new value creation and cultural change globally. Ethically and aesthetically, the social reality experienced through appearances is related to postcolonial times.

Nevertheless, the historical moment that motivated technological development the most in the sense of what today is understood as the technicization of aesthetics is related to the practice of design, in the relationship of applied arts and crafts, which gradually began to approach the Industrial Revolution and consumer culture. The main event of this historical moment was The Great Exhibition of the Works of Industry of All Nations, held in 1851 at the Crystal Palace in Hyde Park, London (1851). The history of exhibitions during the second half of the 19th century simultaneously constitutes a history of iron construction. The exhibition buildings were planned for rapid erection and dismantling, both of which were facilitated by iron. The iron parts for these buildings could be manufactured in widely separated workshops. Iron was employed everywhere as the most appropriate medium of expression. This international exhibition in London became a testing ground for new methods of many other innovations (Hitchcock 1976).

Therefore, technology has conducted the technicization of social structures, building interaction between nature and culture for all possible achievements of human civilization over time. Such achievement defines the spirit of the time (*Zeitgeist*), as evidenced by the ideas and materializations of each context. According to Giedion, this context divided architecture between art and technology. An analysis of the configurations of human social life throughout the evolutionary history of humanity yields the observation, according to Picon (2005–2006), that in the 19th century, the separate existence of an Ecole des Beaux-Arts and an Ecole Polytechnique in Paris highlighted a gap between art, architecture, and construction. Consequently, the search for traditional values, as opposed to the ideal of innovation, in a dialectic constructs narratives for an aesthetic value judgment toward an aestheticization of urban spaces. Architecture in public space has several ways of being presented; for example, it can be figurative and propose its configuration of public space with monuments and areas of socialization, testifying to the importance of architecture's relationship with the great social and material activities of human life. Le Corbusier confirmed that construction is made to hold, whereas architecture is made to move (Giedion 1978). The relationships of architectural works with painting, literature, music, theater, and cinema offer the sensitive aspects of a representation as an experience of individuals in their urban environment. These sensitive aspects relate to aesthetic values, not only in the art of building but as human achievement in art and the aesthetic object.

Throughout life in society, humans have acquired and created countless symbols to give meaning to their experiences and behaviors in various cultures. As a result, humans become accepted individuals in a particular cultural, community, or social environment. The individual lives and gives meaning to life within the organized systems led by cultural patterns. Hence, architecture has transformed, in its history, the way in which it is understood. Whatever the metropolis, architecture's evolution and cultural importance have contributed to its social representation as well as aesthetic and ethical aspects being valued. From prehistory to contemporary society, the transformation of the ways of life has authorized an immense inventory of all of the achievements that cannot be presented here. Therefore, in general, the configuration of contemporary architecture in its quest for innovation can be represented by understanding the dynamics of global metropolises in their transformation process, continuously seeking to meet the needs of the modus vivendi. Architecture finds itself, *vis-à-vis* technological and scientific evolution, in a framework of contemporary universality, which is not only that of social transformation but also that of the environment. In this direction, the essential project lies in developing smart cities.

Smart Cities and Visual Configurations

The urban landscape's transformation is perceived to have resulted from the search for adaptations to an intelligent environment. The challenge in this sense is to find the optimal solutions in architecture and urban design through new technologies that enable sustainability. Concerns with new urban projects are focused on the interaction of daily life in the metropolis of collective life. To this end, new technologies enable new architectural spatial models, creating realities, intelligent environments, and interactive communities. Thus, the artificial environment and the configuration of urban and digital environments are discussed in this section concerning changes in the habits of residents. The following questions are addressed: How is a public urban space that is driven by collective and democratic intelligence possible? Which ethical and aesthetic experiences are possible in a metropolitan city? What is the role of AI as a solution? A smart city is the concept that most closely matches this analysis of the technicization and aestheticization of social and urban structures. For a city to be smart, the interaction of its inhabitants is necessary in terms of access to energy, materials, services, and financing aimed at economic development and quality of life. For this purpose, the following are essential: Human capital; the coexistence of groups of people with different incomes, cultures, ages, and professions; the promotion of economic development with local plans for industry, innovation, and entrepreneurial initiatives; sustainable development; and city mobility. However, these conditions are not restricted to a city as a closed structure that is limited to its territorial space but exist in its global connectivity.

Consequently, communication and information technologies play the leading role in the challenges of a smart city. In addition, digital technologies generally provide new experiences that connect urban and natural environments. Technology has reached a stage where many of the aspects that were once considered negative for humanity are now considered forms of solutions and thus viewed positively for achieving sustainability goals. However, social problems still need to be solved. Many of the visualizations for a smart city are still only in projects, and their implementation still depends on other priorities being solved. The most essential is that everyone has access to new technologies. Much political will and many investments are still required to make Internet access feasible. Digital inclusion can thus become a basic need over time, after health, education, and quality of life.

Crucial aspects throughout development have already been discussed, and they are part of the narratives of everyday urban life and the utopias that shape the uncertain future. However, it is worth remembering that utopias are essential because they represent the idea of a critique of reality, imbued in which is the imagination of an ideal. Therefore, utopia is not simply a word that evokes a kind of social fiction or a type of illusion; rather, the idea of a utopia arises from a critique of social conditions. Embedded in the principle of utopia is critical intentionality, as was demonstrated well by Erika Naginski from the Harvard University Graduate School of Design when she discussed Claude-Nicolas designing what he imagined, namely a utopia he called the *Ideal City of Chaux*, and also Sir Thomas More's novel *Utopia* (1516). More's *Utopia* was published in Latin and later reprinted and translated into other languages. An essential aspect of this work even today is the binary structure. The first part is a critique of contemporary society in England, of kingship as a model for the state and as a political principle. The second part is about a traveler who describes the Island of Utopia. In this binary structure, the idea of a critique of reality as a utopia is developed. The term "utopia" is used for the first time in this novel, which etymologically has two meanings. Derived from Greek, it can mean both "nowhere" and "happy place," thus emphasizing according to Naginski, a Harvard architect, throughout the course *The Architectural Imagination* (2017), that a utopia is not necessarily something unrealizable but rather something that establishes conditions of possibility, whether architectural, urban, in design, or in terms of economic and social policies.

Therefore, through the configuration of the urban image in its process of aestheticization and technicization in contemporaneity, designating time and its relationship with space, one can think about all of these utopias for the future of cities. That is, critically and in the sense of the evolution and perception, think about the new interventions related to communication technologies in sync with the urban rhythm, enabling new dimensions of spaces and relations for a future based on sustainable industry and fair trade. For example, the immediate need for the resources of digital communication and information technologies, for solutions in events and distance learning as well as other remote activities related to work and daily life, is no longer just an idea. In practice, people experienced the use of communication and information technologies as a solution for reducing the spread of COVID-19, which led to a significant reduction in greenhouse gas emissions. These are the first signs that everyone depends heavily on evolution and access to new technologies—not only for mobility but also for remote activity. Hence, these are new habits that result from other conditions of life in society. The problems of our civilization are consequences of this complexity, which is due to technical, economic, and industrial development, involving pollution, global warming, and all threats to the human species. These factors imply increasing needs for production and consumption, especially in the contemporary era where energy resources are saturated due to constant consumption. Therefore, from the end of the 20th century to the present day, the theme in vogue has led the world to project new forms of energy use and a new orientation of society toward a civilization of the 21st century. The 21st session of the annual Conference of the Parties (COP21) set the goal of achieving solutions by configuring a "green image" through sustainability, which involves a shift in industry and consumption.

William Morris was one of the precursors of relating technological and industrial development with the role of arts, design, and architecture to preserve the environment through natural resources, as he was aware of the long-term and irreversible effects of technological progress (Herbert 1992). Thus, the transformations impact the spaces and the population's habitat, and their causes are varied and distinct in each city and culture. Nevertheless, development is always constant at innumerable levels. Otherwise, there could be no expectation of civilization. The relations between industry and city are transformed, evidently, by technological innovation, but these changes are always conditioned by choices. Therefore, knowing the reasons for investments for decisions is mainly a strategic question and rarely of an evident determinism. For example, industries originated in capital cities, where high-quality craftsmanship was found. Later, they were moved to more spacious locations to favor production using manufacturing lines. Port cities had special prominence in this process as they had industrial facilities for exports. Consequently, the development of urbanization is an effect of economic growth and related to changes in the conditions of collective existence.

However, when one addresses visualizations of urban space and the technicization and aestheticization of social and urban structures, one must also consider the role of visual communication and signalization. Historically, already in the Middle Ages the city of Paris highlighted the visual configuration as an orientation for people in urban space to identify streets and houses. The street offers visual aspects and direction in the urban spatial composition through the architectural characteristics of housing and commercial buildings as well as its decorative elements, such as paintings, sculptures, and landscaping that integrates nature with urban space. These distinctions are seen as a play on the notions of zoning and segregation (Leguay 1984, 8). Then, with the emergence of the subway and the automobile, the transformations of public space and collective life followed the consequences of urbanization focused on the development of a signaling system. This signaling system and visual communication development came from private initiatives aimed at advertising in partnership with public institutions. Seeking the organization and maintenance of urban space, this cooperation between the private and public spheres has provided urban furniture essential to public life, such as garbage bins, street clocks, bus stop shelters, and advertising space. On vehicle transit roads, slots are reserved exclusively for messages on traffic signs.

Specifically in the case of São Paulo, a municipal law known as Clean City Law (Lei Cidade Limpa) has directed items of street furniture as a category of the city's information and urban communication. The law prohibits billboard and banner advertisements and regulates totems and commercial signs to reduce visual pollution, aiming to organize and valorize public space.

Therefore, the professor and urban architect Issao Minami (2001) said that it has become increasingly necessary for projects focused on media in urban spaces to solve some problems of "visual pollution," as in the case of São Paulo city. Consequently, artistic interventions in architectural projects have become more valued. Furthermore, the fine arts are essential in organizing visual communication and how we think about the city, accordingly to the architect and artist Claudio Tozzi (Moreno 2002), as Figure 8.4 shows.

The urban signage system is a communication system with its own language and codes. It is an international language of public spaces in large cities due to an empirical development process focused on functionality rather than aesthetics.

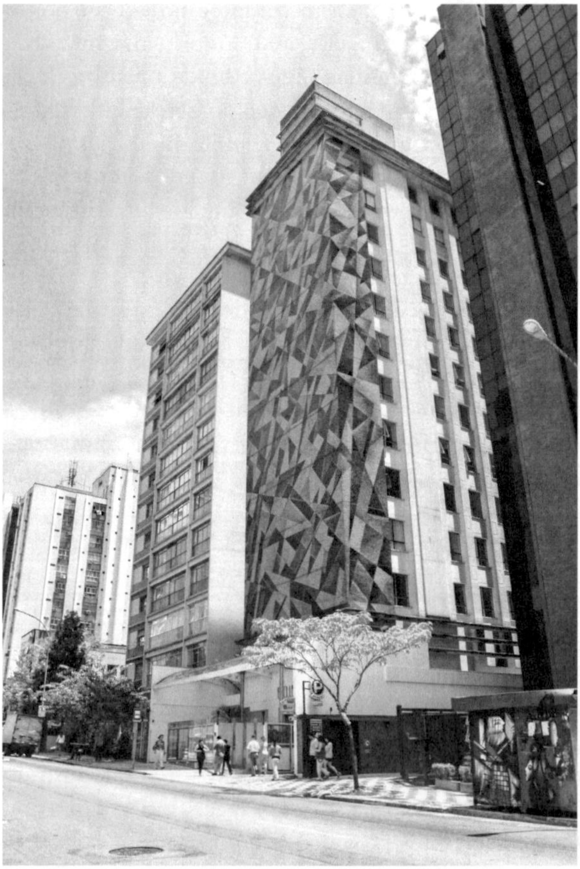

FIGURE 8.4 Claudio Tozzi. *Exclusive Building Mural.* Angelica Avenue, São Paulo City. © Aguina | Dreamstime.

In addition to these technical and functional communication factors in urban space, other contemporary values concentrate on aesthetic and cultural aspects, aimed at a technicization that is focused on the organization of public structure. In 1972, the United Nations Educational, Scientific and Cultural Organization (UNESCO) in its General Conference approved the Convention Concerning the Protection of the World Cultural and Natural Heritage, which identifies and protects cultural and natural sites considered to be of outstanding universal value throughout the world. Today, the Convention is UNESCO's most widely disseminated and visible international instrument, encouraging countries to commit to the protection and enhancement of their properties inscribed on the so-called World Heritage List. Moreover, beyond maintaining the conservation of cultural and natural heritage sites, UNESCO seeks to enhance building capabilities and improve communication. Cities and places on the World Heritage List assume responsibility for protecting world heritage for present and future generations. To this end, the principles of sustainable development are necessarily part of the conservation process, relying mainly on communication tools and strategies in their forms to raise awareness and knowledge of the World Heritage objectives.[4] Each World Heritage property communicates the values of each culture and its identity.

The benefits brought through the recognition of being a World Heritage extend from social, educational, cultural, and political advances to the valuation of the assets in the areas declared World Heritage sites. Among the advantages are contributions from private investors, increased in tourism, employment and income generation, and a more significant international projection. In addition to the aforementioned notions, which are necessary for urban organization concerning visual aspects, the focus is on inhabitants' perceptions of their urban environments and how they orient themselves and behave in the face of everyday experiences. To this end, regarding a smart city's contemporary goals, one must primarily consider the interactions in urban centers through digital technology. However, urban interventions through digital technology must have a social meaning in their applications in visual settings, conditioning the aesthetic aspects as new concepts and, mainly, the functionality to serve everyone through a technicization of the organizational and implementation processes. In this sense, image configuration has practical and empirical significance in daily life in a society, specifically in the visual project, due to its importance to solutions for communicative and signaling systems in large cities. We know that the visual project is essential to social representation today for communication with focused objectives, interrelations, and expected results. To synthesize, the understanding of the condition of cultural worldliness is indispensable to any reflection on the destiny of world cultures in the face of globalization and the flow of industrialized cultural goods, with attention paid to the development of new forms and functions as the specificity of innovation in the configuration of the contemporary image in large metropolises. In this sense, the way of life in urban society today exhibits a standardized cosmopolitan visual layout that involves the aestheticization of urban structures.

The Digital Age and the (Dis)Appropriation of Urban Space

This interdisciplinary analysis highlights the values and effects of contemporary visual culture related to the connected fragments and digital technologies between the physical and virtual spaces regarding the concrete actions of appropriating and disappropriating the urban space. Transformation is the main characteristic of urban visual culture, representing a world conception with features and needs that vary according to culture, society, and economy. Creativity is presented in its realization processes as the meaning of the urban image in its current transformation process, and the configurations of images represent the reality of society. Thus, through visual culture, the action of the present develops around elements such as the economy, techniques, science, ideologies, and political will. These actions transform evolution and, within such a perspective, one can reflect on the conditions that allow innovation and creation in the face of the polysemy of images in this new sense, which is based on the media spreading the ideas and images of mass society. The copious production of images is possible in a context favored by sociocultural dynamics. In other words, this production level is possible because contemporary metropolises and societies report and represent their histories, experiences, expectations, and, in short, their daily lives through images that propose a projection of social identity. This projection transmits, for most people, the collective imaginary. Furthermore, the process of social appropriation[5] of urban space is intended to organize societies according to the needs of each social segment (Lefebvre 1968). The structuring of space depends on the objectives of the public sphere of the city or community in question. The urban morphology is configured and structured by the interdependent elements of the relationship with the public and private sectors, originating spaces related to the city's history. These urban spaces provide social relations through human activities, such as parks, street markets, shopping streets, and areas for cultural activities, such as music, theater, live shows, and art exhibitions.

However, differences exist in the appropriation of the space of collective use concerning cultural, ethnic, and class differences, defining a hierarchy of spaces. In some structures, the hierarchy of space presents more assembled social-relation situations, such as places and passages that everyone must frequent. In terms of quality in their construction forms, these spaces display natural environmental conditions, and landscapes delimit their appropriation. Moreover, with some variations in the state of conservation of the buildings and accessible areas, such as gardens and parks, the urban structure is formed by the activities and presence of its inhabitants. Thus, the urban structure depends much more on the interrelationship of the population with the urban elements than on the configuration of these urban elements without daily life. This interrelationship between inhabitants, architecture, and urban design is a determinant of the appropriation of public space. Therefore, when considering the digitalization of public space, in contrast to the characteristics of appropriation, the term "disappropriation" is used to mean physical space being replaced by digital space. However, from a legal point of view, "disappropriation" is one of the most incisive forms of state intervention in private property, although it can even fall on public property. It is second only to confiscation—for which some reserve the term "expropriation." Nevertheless, in this section, the term "disappropriation" is used to refer to the experience of the empty urban space, dispossessed of its public life by digital platforms. In this discussion, the disappropriation of public space alludes to its urban structures ceasing to be social spaces for daily life. Recently, we experienced this dispossession as many cities were forced into a state of lockdown to reduce the spread of COVID-19. The world functioned virtually, and social relations occurred exclusively through digital communication. Public spaces in cities became empty and lifeless. On the one hand, due to needs of public health, we conditioned ourselves to a reality of the disappropriation of public space and transferred our activities to virtual platforms. On the other hand, those who did not and do not have access to digital technologies became conditioned to isolation and marginalization. This situation is a reality evident even under normal conditions; considering the growth of virtual activities that we have all experienced, no empirical research is required to prove the results. Therefore, some research results demonstrate that "the digital revolution would no longer only be industrial but also fundamentally political: it would redistribute power in urban planning and urban governance," according to Languillon-Aussel (2021) when discussing public space studies. The author argued that "many researchers have already identified in other fields . . . the political economy of markets regulated by public power (such as taxis or hotels)." Therefore, this experience is part of our everyday lives. Nevertheless, this reference to a disappropriation of public space is primarily related to public health needs due to the pandemic context. However, the condition of this physical public space having migrated to digital space is seen as an extension of public space and a trend of human behaviors for the reasons already discussed in "Extending the Public Image on the Web" and "Deterritorialized Spaces" (first chapter sections).

Therefore, the digitalization process proposed by the Fourth Revolution in projects aimed at a smart city consists of digital technologies in production processes, education, and services for optimizing and making processes more efficient. Furthermore, the integration of computer systems works in a coordinated manner to exchange information between different systems. This connectivity is where the IoT stands out, for the interconnection between objects enables infrastructure (electronics, software, and sensors) organized in networks, thereby facilitating communication and interaction. These are the main features that outline the Fourth Revolution and are part of the projects aimed at a smart city, essentially including cyber security. This security is inevitable and concerns hardware and software infrastructures for protecting information that is processed, stored, and transported by digital information systems. However, in connection with the physical urban space, security is also a concern, especially in terms of increasing violence in large metropolises. For this, surveillance and monitoring services using new digital technologies, through security cameras, are employed to provide greater security, including the use of facial recognition to secure access to specific locations. Through technological integration, cameras with remote connections transfer images in real time to servers in the cloud (datacenters), which can be accessed from any device connected to the network. The right to security guarantees that public authorities can install this technology to monitor public spaces. However, other rights are at issue when these forms of monitoring come to be used for control—when public power is exercised in an inappropriate, misappropriated manner, it is a matter of the right to privacy in public space, as already discussed in "Urban Space and Collective Consciousness" (Chapter 2).

Consequently, statistical and machine-learning techniques are required to solve problems with data of greater variety and complexity, namely big data, which arrive in growing volumes and at an ever-increasing speed, making them impossible for humans to solve. In addition, to the greatest challenge of improving human qualities, AI offers an advanced analysis based on logical techniques and machine learning. This enables situations to be interpreted and trends and system behaviors analyzed to support, automate, and make decisions. Now, where is all of this public life in the physical urban space that has moved to the virtual? The answer is in the cloud. The cloud is not something in the atmosphere but rather the distribution of computing services—it is a physical place. It is a global network of remote servers for storage, databases, networking, software, analytics, and intelligence around the globe that are connected and operate as a single ecosystem. They are connected by the Internet with the use of memory, storage capacity, and computers, providing flexible resources and economies of scale. Finally, one must consider the other digital technologies for this world appropriated by digital technology, such as robotics, integrated systems, computer-based technologies that enable 3D simulation and visualization, and even 3D modeling software and 3D printers, which produce parts from a digital drawing.

Thus, what is the main objective of this fourth revolution? What are the benefits of the consequences of a possible "disappropriation" of the physical urban space? The answer is simple and is part of the inference made at the beginning of this discussion—profits, the fourth phase of the Industrial Revolution and the foundations of financial capitalism. For example, in research conducted between May 2018 and October 2019 by the Brazilian National Confederation of Industry (CNI), of which the results are emerging through a pilot program named More Advanced Industry (Indústria Mais Avançada), the impact on market production was found with low-cost tools such as sensing, cloud computing, and the IoT, highlighting that the benefits achieved with Industry 4.0 are numerous. Hence, digital technologies in the industry have allowed an average increase of 22% in production capacity (CNI 2022).

Therefore, distinguishing the political and economic aspects from the social and urban structures in their immediate needs to find solutions for the socio-spatial environment is not simple. This chapter has discussed technicization and aestheticization under the historical achievements of practical and aesthetic developments considering urban space—physical or digital—through the visual aspects of the entire environment. On the one hand, the discussion has addressed the achievements that are part of the cultural development within territorial limits—appropriating the physical urban space. On the other hand, the discussion has addressed concerns about deterritorialization, achievements related to space-time, and images moving in the face of the perceived image—disappropriating the physical urban space. Now that these points have been analyzed, Chapter 9, titled "The Urban Image in Motion and Connection," addresses the formulation of an enhanced understanding of the complexity of urban lives. This is because the urban space is one of the primary environments where artistic and cultural achievements, integration, social inclusion, and new visualizations are manifested.

Notes

1 According to *The Oxford Dictionary*, the term dates back to the 1930s (from technicize + -action). The term means the action or process of making (increasingly) technical or technological in character.
2 According to *The Oxford Dictionary*, the term dates back to the early 20th century; the earliest use was found in *The Jewish Quarterly Review* (from aestheticize + -ation). The term means the action or process of making something aesthetic in character or appearance; an instance of this.
3 Barry Sanywell:

> A configuration of images, archetypes or image-concepts institutionalized in a given culture and society. For example, modern cultures that have institutionalized and externalized the production, distribution and reproduction of the imaginary (the socialization and commercialization of Kant's *Einbildungskraft*). In contemporary mass-mediated culture, image-production takes the form of mass-produced imaginary formations leading to the aestheticization of politics that Walter Benjamin famously associated with the propaganda apparatuses of German National Socialism or the outpourings of culture industries empowered with the means of global mechanical (and now electronic) reproduction. These modern imaginaries are typically projected as objects of desire and emulation (for example, the Hollywood institution of the 'star system' and the obsession with conspicuous heroes and celebrities in contemporary society).
>
> *Dictionary of Visual Discourse* (Burlington: Ashgate, 2011), 371

4 See the Future of the World Heritage Convention. Paris, UNESCO Headquarters, 7–9 November 2011. https://whc.unesco.org/archive/2011/whc11-18ga-11-en.pdf.
5 About the concept of the appropriation of public space, the work of Henri Lefebvre (1968) stands out. The concept of appropriation means:

> Citizenship is defined to include all urban inhabitants, conferring two central rights—the right to participation and to appropriation. Participation allows urban inhabitants to access decisions that produce urban space. Appropriation includes the right to access, occupy and use space, and create new space that meets people's needs (Lefebvre 1968 in Kofman and Lebas 1996: 174). Lefebvre was writing just before the 1968 students riots in Paris, and his ideas won popular acclaim. Of importance was his emphasis on the right to the city as a whole, rather than to specific rights in cities, but his writing was disconcertingly vague as to how it could be implemented.
>
> Alison Brown and Annali Kristiansen, "UN-HABITAT project on Urban Policies and the Right to the City: Rights, responsibilities and citizenship" (March 2009) UNESCO. https://unesdoc.unesco.org/ark:/48223/pf0000178090

Bibliography

Benjamin, Walter. "Das Kunstwerk im Zeitalter seiner technischen Reproduzierbarkeit" [The Work of Art in the Age of Mechanical Reproduction]. 1939. Reprint, *Medienästhetische Schriften*. Frankfurt am Main: Suhrkamp, 2002.

Brazilian National Confederation of Industry. "Working to Develop Industry and Brazil." Last modified April 17, 2022. https://www.portaldaindustria.com.br/cni/en/

Giedion, Siegfried. *Espace, temps, architecture* [Space, Time, Architecture]. Paris: Denoël-Gonthier, 1978.

Giedion, Siegfried. *Architecture et vie collective* [Architecture and Community Life]. Paris: Denoël-Gonthier, 1980.

Giedion, Siegfried. *Construire en France, construire en fer, construire en béton* [Building in France, Building in Iron, Building in Concrete]. Paris: Éditions de la Villette, 2000.

Herbert, Isolde Karen. "Nature and Art: Morris's Conception of Progress." *Journal of William Morris Society* v. 10, no. 1 (Autumn 1992): 4–9. https://morrissociety.org/document/vol-10-no-1-p-04-09-nature-and-art-morriss-conception-of-progress/

Hitchcock, Henry-Russell. "The Crystal Palace: Ferrovitreous Triumph and Ensuing Reaction." *Early Victorian Architecture in Britain*. Boston, MA: Da Capo Press, 1976.

Languillon-Aussel, Raphaël. "Digitalization of Public Spaces. The Great Urban Change?" *Journal of Urban Research* 22 (December 2021): 1–20. https://doi.org/10.4000/articulo.4859

Lara-Hernandez, J.A., Alessandro Melis, and Steffen Lehmann, S. "Temporary Appropriation of Public Space as an Emergence Assemblage for the Future Urban Landscape: The Case of Mexico City." *Future Cities and Environment* 5 (2019): 1–22. http://doi.org/10.5334/fce.53

Lefebvre, Henri. *Right to the City*. 1968. Trans. by Kofman, E. and Lebas, E., eds., *Writings on Cities*. Oxford: Blackwell, 1996.

Leguay, Jean-Pierre. *La rue au Moyen Âge* [The Street in the Middle Ages]. Rennes: Ouest France, 1984.

McLuhan, Marshall. *Understanding Media: The Extensions of Man*. New York: McGraw-Hill, 1964.

Minami, Issao. "Paisagem urbana de São Paulo: publicidade externa e poluição visual" [São Paulo's Urban Landscape: Outdoor Advertising and Visual Pollution]. *Vitruvius*, June 2, 2001. https://vitruvius.com.br/revistas/read/arquitextos/02.013/879

More, Thomas. *Utopia*. 1516. Trans. by John P. Dolan. In *The Essential Thomas More*, edited by James J. Greene and John P. Dolan. New York: New American Library, 1967.

Moreno, Leila K. "A Angélica é azul" [Angelica Is Blue]. *Jornal da USP*, September 1, 2002. http://www.usp.br/jorusp/arquivo/2002/jusp610/pag1011.htm

Naginski, Erika. "Ledoux's Utopian City of Chaux." The Architectural Imagination. Course at the Harvard University Graduate School of Design, 2017.

Picon, Antoine. "Construction History: Between Technological and Cultural History." *Construction History* 21, no. 21 (2005–2006): 5–19. http://www.constructionhistory.co.uk/shop/

Sanywell, Barry. *Dictionary of Visual Discourse*. Burlington, VT: Ashgate, 2011.

Welsch, Wolfgang. "Aestheticization Processes: Phenomena, Distinctions and Prospects." *Theory, Culture & Society* 13, no. 1 (February 1996): 1–24. https://doi.org/10.1177/026327696013001001.

9

THE URBAN IMAGE IN MOTION AND CONNECTION

Architecture, design, and visual arts are, in their conception, the technique in their creative processes, presenting configuration goals related to society's material aspects—that is, economic, scientific, and technological development concentrated in the urban environment. Beyond the medium's importance as well as the form and content in which these arts configure their characteristics, this chapter considers the various forms of visual expression in large cities that meet an aesthetic experience of everyday life. Images are related to cultural manifestations and meaning in urban and architectural space-time. Various visual narratives are experienced in their significance and relationship with movement and synchronization with sounds in everyday interactions, approaching digital technologies. Therefore, the correspondence between architecture and cinema is a good reason to consider the innovation involved in configuring images in the connections between physical and digital urban spaces. Meaning would not be created in the moment of execution but in the visual narratives' formal result, which concerns our perception of time and space as reality. Additionally, meaning would be determined as a reciprocation of the perceptive relations between creative individuals, artists, architects, designers, filmmakers, videomakers, and image manipulators (Chapter 2). In addition, based on Nouvel's experience as an architect, the notions such as displacement, speed, and memory are cognitive domains. Furthermore, these concepts are compelling aspects of filmmaking. These ideas in architecture relate to how one moves in an environment of architectural space (Baudrillard and Nouvel 2000). Hence, these perceptions are shaped by the interplay between the present and the time of the visual narratives' creation. The insights are also shaped by how the creations of the architect, artist, filmmaker, videomaker, or image manipulator influence this interplay.

DOI: 10.4324/9781003314943-10

Based on the aforementioned considerations, the purpose of this chapter is to approach the technical effects of the moving image and architectural structures in the interaction of physical and virtual spaces through digital technologies. Consequently, public and democratic values are analyzed considering the use of surveillance cameras and the connection between the physical and virtual spaces that they enable. The camera represents the main element in this connection and interaction through capturing images of the human presence in its activities. On the one hand, this analysis is about capturing the image. On the other hand, it is about the necessary conditions for this capture concerning the limits of the rights exercised by the state as a public power or by the private sector in its rights to exercise public access practices. This approach to the moving image and digital connection is directed to the visual experience in physical and virtual urban spaces based on the objectives of a smart city's configuration (Batty 2013; Picon 2015; Picon and Hill 2020).

However, the smart city configuration is conditioned to the surveillance and digitalization systems of all segments of static and moving physical spaces. More than half of the world's population lives in urban areas.[1] In urban spaces, several architectural structures integrate screens. Screens are spread everywhere in cities, transmitting recorded and live images connected to various channels, applications, and websites in a complex user experience through hypermedia and multimedia systems. They directly connect physical and digital spaces, thus connecting urban life with citizens. Capturing and transferring these images is made possible by cameras, through which the focus is directed to every detail of urban space. Captured images undergo transfer, encoding, and storage. This process is a determinant in the effects and manipulations that are possible using digital technologies to integrate spaces and times, enabling individuals to be active and passive forms of visual and sound participation. Thus, media are the primary tools of intermediation between sender and receiver. A digital interface, tablet, smartphone, computer, or other device that transmits media digitally allows visual and audio access, as Figure 9.1 shows. According to the Digital 2022 Global Overview Report,[2] the number of users who own one of these devices—of which the smartphone is the most common—is equal to more than two-thirds of the world's population.

Even though ideals, novelties, and technological advances stimulate theories, discussions, and critical analyses about the near future of urban space, such implementations are in practice still unfeasible. While the future is glimpsed, the past remains in coexistence, seeking to overcome the challenges of updating technologies. Moreover, many countries still face challenges related to technological upgrading that depend on economic factors and fundamental social solutions. Consequently, the following question arises: Under what conditions would urban projects offer digital solutions for society? For enhanced clarification in search of an answer, this chapter's analysis is directed at digital technology in its specific coding and optimization of space-time concerning the interests of society and, consequently, in its technical and economic effects on urban structures

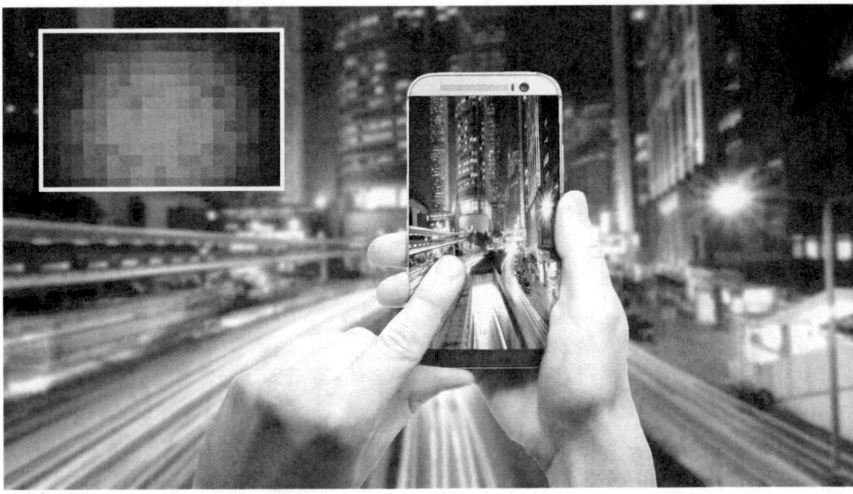

FIGURE 9.1 Image captured using a smartphone. Bitmap image formed by pixels and zoomed-in detail bitmap image shown in 8-bit grayscale.

and social transformation. Thus, in the first section, titled "Digital Technology, Interactivity, and Integration," a brief notion of multimedia and hypermedia systems is presented, followed by the technical aspects of converting from analog to digital media, the still and moving images captured by cameras and how they are visually perceived by the human eye, as well as the possibilities of virtual environments. The goal is to understand—"behind the scenes"—the configuration of urban space between physical and virtual spaces, focused and controlled by cameras in the face of digital resources and the limits of human perception. Next, in the second section, titled "Simulation and Reality," the applications of artificial intelligence (AI) are presented in different simulations and realities for automating various processes that involve the functioning of an urban city. This is a process of simulations becoming local realities with specific focus spaces, thus representing a shift from geographic to geometric representations. In the following sections, "Internet of Things (IoT), Artificial Intelligence (AI), Machine Learning, and Algorithms" and "The City's Quantitative and Qualitative Values," preliminary notions about the technical characteristics of digital resources will be presented. In the final section, titled "Smart Cities, Visual Media, Data Systems, and Communication," the role of new technologies in their social connections with services is analyzed in their behavioral implications regarding everyday experiences in the sense of a smart city.

Digital Technology, Interactivity, and Integration

The development of digital technology and network communication through the World Wide Web implies transformation from the analog formats of films, photographs, and printed images to digital formats and new productions created in a digital format. In this way, the characteristics of each form, such as film, photography, and television, and their unique values as media are maintained but updated with the new digital technology and uploaded to the Internet. The well-known technological convergence occurring through the World Wide Web is currently the most significant result of the digital technological evolution. It is a system that provides the user with simultaneous access to various means of communication, all connected on the same device. Keeping in touch with all other users of the technology directly or indirectly, actively or passively participating, and receiving or emitting information are all possibilities—it all depends on what one chooses from among the numerous possibilities for interaction. With the decreasing costs of digital devices and thus their popularization, it is possible for any individual to film, photograph, and produce sounds and images demonstrating their content and authorship. Of course, as already exhaustively noted throughout the discussions presented in this book, one must also consider that globally, many still do not have access to digital technologies due to social inequalities. However, considering that those who define the meaning of the market possess the technology, an infinitude of independent productions with the most varied repertoires emerges, supported by the facility of their participation on the Internet. The predominant language of this digital landscape is hypertext, which is highly familiar to most Internet users. This term originated with Ted Nelson in the early 1960s, but it was several more years before hypertext was developed by Douglas Engelbart just before the World Wide Web emerged (Bardini 2000). The concept of hypertext was central to the creation of the World Wide Web. Virtually all companies, organizations, and institutions use the Web for various activities, such as managing projects, offering distance learning, giving virtual tours of museums, ordering materials, and distributing company information. The Web provides servers through the open Internet system as well as private and secure servers called an intranet. In this sense, graphical Web browsers offer pictures, sound, and video capabilities to be embedded in the text and applications to be embedded in the multimedia system. Multimedia is dependent on information and communication technologies for direct (nonlinear) access, interactivity, and integration with programs and applications. Its spatiotemporal nature is classified into three media forms. First, there are discrete or static media, such as texts, static images, and graphics. Second, there are media that are continuous or dynamic because their content changes over time, such as audio-video and animation.

Finally, immersive media present interactive information elements in three-dimensional (3D) environments that allow substantial interactivity with the user. As for the degree of linear interaction, the user has no control over the multimedia reproduction process, which differs from nonlinear interaction. The user can interact through navigation. He or she accesses information in a nonlinear and autonomous manner through various means of representation and expressions of content. Moreover, the user can interact freely with multimedia.

The basic classes of multimedia applications and services are conversation, messaging, retrieval through access to multimedia information servers, and broadcasting. Thus, the production of images through the digital option offers an interactive dynamic of fragments of the urban image. This production concerns the diverse images transmitted to the visual culture in forms, formats, and meanings. However, these meanings are also related to the variety of the visual forms of perception, considering how the visual experience is presented. In this regard, among the world of material things, the natural, traditional media, and the virtual, technological transference occurs from the material, natural world and the analog, representing conventional media, to the digital.

In general, the digitalization of information refers to the transformation from analog to digital technology when data from traditional media are converted to digital media. Converting an analog signal to digital is based on quantifying the signal amplitude value at various time instants. In the context of the digital encoding and decoding of digital images, information ignored by the human visual system or that is irrelevant to the intended use of the image is omitted. Therefore, this interdisciplinary analysis of the image is worth highlighting in the perception that concerns the values and effects of contemporary visual culture. Furthermore, the investigation is related to the connected fragments and algorithmic rules between the physical and virtual world in the imagined and concrete realizations of the urban moving image. As for moving images, video compression systems avoid data expansion in two ways. First, they monitor the motion of objects and compensate for it during the prediction and differentiation process. Second, they switch to an alternative coding method when the correlation between frames, through similarity, is insufficient to benefit from predictive coding. Successive frames in a video sequence are often highly similar. Coding their differences can reduce temporal redundancy and enable significant compression. However, the similarity between nearby frames is reduced, and compression is negatively affected when the frames' sequence contains fast-moving objects, zooming and camera spins, sudden changes, or gradual increases and decreases in scene light intensity (fade-in and fade-out, respectively). In most images, pixels are correlated in space and time when the image is part of a video sequence. Since most pixel intensities can be reasonably predicted from the intensities of the adjacent pixels, the amount of information conveyed by a single pixel is small. Much of its visual contribution is redundant because it can be inferred from its neighbors. A two-dimensional (2D) intensity arrangement must be transformed into a more efficient but usually nonvisual representation to reduce the redundancy associated with correlated pixels in space and time. In image coding, algorithms seek to eliminate statistical, spatial, and *psychovisual* redundancies. Therefore, animated images represent a significant advance in the human—computer relationship, as does the digitization of sound. *Psychovisual* redundancy exploits the properties of the human visual system based on greater sensitivity to distortions in soft and dark areas as well as greater sensitivity to changes in luminance than chrominance (Gonzales and Woods 2018, 576–594).

One of the most popular techniques for removing coding redundancy was developed by David Albert Huffman (1952). Huffman's encoding results in the smallest possible number of code symbols per source symbol by individually encoding the symbols of an information source.[3] The sequence of code symbols can be decoded in a single form. To "isolate" the information to which the human eye is less sensitive, one must transform the image from the spatial domain to the domain of *psychovisual* redundancy. In this sense, one must understand the image in its forms of perception in distinct realities; one form concerns the perception of images in the physical world, while another in analog technologies and another in digital. The idea is that the configuration of images presents a distinction through information and communication technology through a specific codification—the Huffman codification—among the dynamics of other codifications in the digital processing of images. Images pass through compression algorithms called codecs, which are capable of performing both encoding and decoding to reduce the amount of data required to represent an image in pixels, as illustrated in Figure 9.1.

These codecs are techniques for compressing and transmitting digital video. For example, the MPEG (Motion Picture Experts Group) family of codecs extends JPEG (Joint Photographic Experts Group)[4] to animated picture sequences according to the coherence between frames. It is an essential digital compression standard. The JPEG encoding procedure is achieved using a variable-length code to eliminate redundant information from an image. The decompression process is the exact inverse of the compression process in terms of the sequence of steps (Gonzales and Woods 2018, 539–553).

In an article titled "Media Aesthetics, Simulation and New Media," Jeans Schröter (2013) emphasized the role of the virtual camera in the simulation process, using digital technology in film productions as a simulation through the process of transferring data from analog to digital photographic media—characteristics of the image that are used as templates for new digital representations. In addition, Schröter demonstrated that using a virtual camera is like an analog camera in approaching the reference model. However, manipulation of the image capture is possible with a virtual camera, unlike physically real cameras. This image capture process occurs in virtual space in dimensions and lighting, offering the same visual appearance as analog photographs. However, Schröter highlighted the distinction of materiality in traditional media by perceived forms, for what he calls "transmedial intermediality" (*Transmediale Intermedialität*) for the process through which forms are mediators of shared visual structures. Materiality in this sense is hidden by the "transmedia" forms and is only perceived when the media presents an imperfection in its configuration. One can thus understand that the central aesthetic characteristic between the digital and analog image lies in the possibilities of imperfections of materiality. In this sense, as these digital simulations involve all of the dimensions of physical and virtual realities, one can realize the importance of the imperfections of analog media for evidencing the materiality of physical reality.

Simulation and Reality

Simulation plays a significant role in all production sectors and is made possible with digital technology. Technically, simulation offers experiences through models that represent reality in various situations. Moreover, digital technology is fundamentally driven by mathematical thinking and, in the case of architectural and urban environments, the basis is dynamic geometry. Nevertheless, models must be validated against experimental data for an efficient simulation. Simulation is a technique that involves imitating situations, various operations, and types of real-world processes and facilities that employ digital technologies. It concerns methods of calculating the effects, consequences, and results of situations that humans would experience by the mathematical logic that underpins digital technology.

A simulated model is developed based on specific parameters and tested for refinement and similarity with physical reality under different conditions. Simulation in three dimensions is essential in the spatial context for location (longitude, latitude, and altitude) and 3D objects (length, breadth, and height). With 3D simulations, an entirely virtual configuration or reproduction with the possibility of interaction through virtual reality is possible, with designs limited to three (where they occur) or four dimensions (space-time). There are also other simulations that involve complex resolutions, combining physical phenomena such as dynamics. Consequently, the combination of 3D and multi-physics simulations offers more accurate results. Through these simulations, it is possible to test a large part of the variables of a process without spending on raw materials and with a considerable reduction in direct labor. This technique is also applied to entire cities, simulating basic sanitation, the power grid, car traffic, and patient flow in a hospital among others. AI makes these processes possible through data from technical tests and simulations virtualized in large proportions with accuracy and are already being incorporated into the software. Furthermore, AI enables complex connections that are not possible for humans or static simulators (Hoppin and Berti 2021).

AI, the automation of intelligent behavior, is divided into the following five paradigms: Connectionist approaches, symbolic approaches, evolutionary computation, swarm intelligence (SI), and ensemble learning. These AI paradigms that differ in application possibilities consider determinism, generalization, pattern recognition, creativity, and autonomy for digital transformation. For example, the digital transformation of entire cities already involves some automation of intelligent behavior with Google Maps. Like smartphones, video and photo cameras are integrated to capture images with high-tech digital capabilities through AI systems, such as the calculation of focus and adequate lighting for image capture. With the symbolic paradigm, AI technology is applied in route calculation applications, such as Google Maps, which perform symbolic transformations (numbers, letters, words, and symbols) with precise encodings; thus, they are employed in health, legal, and tax areas to avoid risks. The AI connectionist paradigm concerns applications ranging from smartphone facial and voice recognition to complex commands in autonomous cars. They are currently used by Google, Facebook, and Tesla in their products and services. For example, by using photographic or spectrometric images in colors invisible to the human eye, it is possible with little effort to create systems that learn the texture, aspect, strength, and flavor of products; moreover, face detection can be used in the monitoring of controlled areas to detect possible break-ins (Santos 2018).

Moreover, AI in its evolutionary paradigm presents algorithms capable of reproducing the process of natural evolution to simulate digital ecosystems. This refers to environments created by numerous interconnected computers in the form of a cluster used in industrial design for product development, following the model of Darwinian natural evolution. As for the SI paradigm, convergence algorithms are based on phenomena in nature and are applied to optimize logistic routes for road freight or complex urban environments. Finally, as the name implies, the ensemble learning paradigm is the joint application of all AI paradigms in futuristic projects, such as those of Tesla and Google. This combination of AI paradigms offers improved determinism, generalization, recognition, creation, and autonomy according to the goals to be created as well as the use of AI over AI itself in defining complex technical parameters to reduce the risk of AI failure (Santos 2018).

Internet of Things (IoT), Artificial Intelligence (AI), Machine Learning, and Algorithms

Given this convergence of digital technologies and without the goal of supplying technical depth, some basic notions about the main concepts of the Internet of Things (IoT), Artificial Intelligence (AI), machine learning, and algorithms are presented here to guide our subject. The IoT is a network of things connected by sensors and other devices that are necessary to enable the exchange of data over an Internet network. In this sense, the interrelation of network devices connected by the Internet allows for the transfer of information without the need for human commands, in the communication of machines with machines, through AI. That is the IoT, platforms that make it possible to connect things remotely as objects send and receive data over the Internet to function according to their designated tasks. However, other capabilities involve the use of AI in machines that can interact like humans with humans. In this scenario, humans and machines work in collaboration.

AI is the possibility through which machines learn to perform a task. In this machine-learning process, algorithms and methodologies enable software to improve its performance as it receives data. However, the machine does not need algorithms produced by humans to perform tasks; machines produce the algorithms. Artificial neural networks (ANNs) are algorithms that mimic the biological structure of the human brain. As such, machine learning is the practice of using algorithms to collect data, learn from the data, and execute the goals determined by humans. Through a large quantity of data and through algorithms, machines acquire the ability to learn how to perform tasks without the involvement of humans. From these conceptions of machine learning and AI practices, collaboration between machines is possible, including remotely through deep learning, as in the case of driverless guided cars and other autonomous operations. There has been substantial progress in machine learning and AI with improvements in perception. As a result, bolder applications have been inspired and consequently involve high risks, such as self-driving cars, robotic surgery, finance, real-time control of the power grid, and autonomous weapon systems. However, Thomas Dietterich (2017, 7) posits the need for AI to evolve to reduce risks: "autonomous cars have the potential to greatly reduce the loss of human lives due to human error. But we know that computer vision systems based on machine learning sometimes make errors that no human would make." Similarly, with regard to surgeries using robotic surgical assistants in the healthcare field, Dietterich points out that "the potential for improved medical outcomes is great, but flaws in perception, reasoning, and execution could place human lives at risk." In this sense, Dietterich (2017, 21) notes that these applications need much more efficient AI systems and that the current ones are not capable enough. He observes that "to create the level of robustness required for such high-risk applications, we need systems that are robust both to known unknowns (the uncertainty they represent explicitly) and to unknown unknowns (unmodeled aspects of the world)."

The City's Quantitative and Qualitative Values

City planning is in constant motion and has been intensified in recent decades through digital resources. One of the features of digital technologies is the IoT, which makes it possible to connect a diversity of services online, enabling interaction between things and, of course, interaction with the user. Everything is connected. This reality of the IoT connects devices through digital technology and integrating services in urban and private spaces, homes, household appliances, and public services. In this sense, the IoT and AI are collaborating on planning smart cities to improve city management and offer the population a better quality of life through public services. For example, in São Paulo and other Brazilian cities, sensors installed in public spaces enable several services to function better, improving the planning of public transportation schedules, and storing traffic information. These sensors are the basis for the functioning of some mobile applications using the Global Positioning System (GPS). GPS allows intelligent routes to be drawn for some essential public services such as collecting the garbage, facilitating and optimizing the work of the police, tracking stolen vehicles, detecting irregular cars, and facilitating and optimizing the work of the police, according to information from the Ministry of Communications for the Brazilian government. The concept of smart cities has been the focus of urban planning and the re-urbanization of many cities. However, some obstacles still need to be overcome regarding investments in these advanced technologies and in the digital implementation that allows communication between different systems and easy access to data and services. For example, in Brazil, investments in this technology, according to the Ministry of Communications, are already a reality with the implementation of a new generation of wireless Internet access technology by the National Telecommunications Agency (Anatel) so that all Brazilian capitals have this technology.

An essential aspect of urban design is the balance between the market and the economic context and creating a new relationship between public infrastructure and its place in cities, as "digital technologies create new and challenging opportunities in the field of sustainable urban design," as stated by Friedman (2021, 139). Therefore, in the face of this new digital reality, human life follows the new path on which all things are under the control of the digital age, IoT, Big Data, and AI; in short, under binary reading and algorithms. Thus, to think about the city is to think about all the conditions in which urban life is inserted and, as a basis, the functional and aesthetic aspects of urban and architectural planning. In a recent article, "Is the city becoming computable?", Picon and Hill (2020) present positions on urban planning based on tracing the history of representing cities in the process of digitalization by clarifying that since European urbanization in the 19th century, this topic has been discussed in academia in different ways. They state that there is a romantic point of view, "a belief in the city as fundamentally beyond the powers of human comprehension and thus best apprehended through qualitative means," while opposite this view is the tradition of quantitative values, "a belief that the structures, patterns, and dynamics of cities can be reduced to numbers" (2020, 29). Well, would this reduction to numbers precisely represent the city, that is, as mathematically constructed? Contrarily, the authors give significant consideration to understanding that the city is a social system involving the time-space relationship, while clarifying that "many features that seem to define cities—the abundance of creativity, cultural richness, the fast pace of life, etc.—have been beyond the traditional methods of scientific analysis, precisely because these parameters themselves elude exact definition or measurement" (30). According to Picon and Hill, compared with qualitative values, quantitative values are essential for evaluations in a science of cities. This city science originated in the 19th century with urban planning. The urban planning concern arose with the rapid growth of cities, the problems resulting from this growth, and the need for control and organization aimed at obtaining quality of life. It is in this context that the first statistical agencies appeared in Europe for the quantitative analysis of mortality, birth, crime, and essential conditions of sanitation, education, and public health. Then, beginning in the second half of the 19th century, a profusion of data analysis and visualization methods appeared in Europe and North America. The development of a science of the city emerged effectively at the end of the 19th century based on quantitative social thinking as spatial and, according to Picon and Hill, "only in these closing decades did urban and territorial concerns come center stage, and only then did a set of techniques develop to analyze the urban and geographic aspects of social data" (2020, 31). Historically, it is evident that the development of quantitative scientific methods for analyzing cities resulted from the need to reorganize the structure of the urban environment (33). Another stage that marked the development of cities under the control of the science of numbers was the emergence of cybernetics after the Second World War, and "in the middle decades of the 20th century, digital computation seemed to offer the possibility of bridging the divide between the social and natural sciences—of putting the study of society on a solidly scientific basis" (35). One of the main reasons for the technological advances and the control of city flows and spaces, especially in North American cities, was the insecurity generated during the Cold War, which has been extended to the current time. This insecurity is due, on the one hand, to the Cold War period and avoiding any possibility of atomic attack and, more recently, against any case of terrorist attack following the September 11, 2001 event. On the other hand, every new technology is tested in secrecy, utilized by the state intelligence services, and used in cases of defense or attack, as discussed earlier (see Chapter 7). Furthermore, cybernetics offered no solutions for city planning. The basics of cybernetics applied to the context of cities did not consider the primary forms of human internal and external perceptions of time and space, that is, "the facts that historical circumstances matter, that past events, and decisions constrain future developments, that geographical unevenness is both cause and consequence of social and economic dynamics" (Picon and Hill 2020, 36). Thus, the authors clarify that the critical highlight of advances in communication and information technologies arrived with the discoveries of Jay Forrester, engineer and researcher at the Massachusetts Institute of Technology (MIT). However, his research was focused on the development of state defense technologies, leading to the design of the SAGE missile defense system in the 1950s, which, later in 1961, served to address communication and optimization issues in human and machine network systems. Forrester has developed complex system dynamics to apply the new mathematical and computational methods corpus (Picon and Hill 2020, 35). Cybernetics then consolidated the science of the city under the control of numbers in Forrester's *Urban Dynamics* (1969) as the foundation for a computational structure to simulate how cities function. The authors clarify that "Forrester devised a set of equations describing how all of the major components of urban systems operated and interacted with one another" (35). However, with the current advances in research, complex digital systems now make it possible to simulate social phenomena involving time and space. Furthermore, Big Data is not specifically a feature attributable to the smart city concept but to computational methods and traditional modes of data analysis. According to Picon and Hill, Big Data "is more a product of the ubiquity of the smartphone, and thus should be dated to slightly after the dawn of the smart city" (39). In general, the logic of Big Data is the ability to quantify all aspects of urban life by numerically representing reality and archiving it in the cloud.

In the technological evolution of the individual and their know-how, humanity follows, without control, a romantic journey of dreams and achievements on an invented artificially built path as everything materializes, while adapting, growing, and absorbing all technology without knowing why. A synesthetic relationship has been established between human beings (users) and technology (products). It is the materialization of ideas as a consolation for overcoming nature, characterized throughout the evolution of technologies as a way to better relate to environment, even under the pretentious control of some social and ideological systems. However, as previously discussed in the chapters "Cultural Images: Real and Imagined Experiences" (Chapter 1), "Urban Space and Collective Consciousness" (Chapter 2), and "Sociopolitical Facts, Art, and Technology" (Chapter 7), the possibility of technology dominating humanity is not recognized. It is nonexistent, because the acceptance of this condition is not compatible with human behavior but rather the use of technology by some humans to increase control or domination over other humans, as discussed by Norbert Wiener (1950), creator of cybernetics. Thus, the good use of city science through new technologies and quantitative analysis methodologies offers improved safety, security, monitoring, and efficient mobility or its misuse to watch, invade, disrespect, and track the possibility of displacement.

Smart Cities, Visual Media, Data Systems, and Communication

Thus, information and communication processing has migrated from the newly learned technical domains to the new media. As Marshall McLuhan stated in *Understanding Media: The Extensions of Man* (1964), new forms of media are always integrated. Therefore, several levels of interaction with information and a diversity of purposes must be considered when using the Internet. In this regard, the role of the media and knowledge versus information should be highlighted. According to Pierre Lévy (1990), the third form of knowledge appropriation falls in the space of new communication and information technologies. There are many technologies, and their purposes and functions are diverse. The speed of technological advances can be verified by the unfolding diversity of technologies in recent years. Many other uses of digital equipment have also become possible with the articulation of computers in networks. Thus, the multitude of new technologies presents a variety of possibilities for action and communication. Therefore, knowledge construction is sought in the information process, and the media are the primary intermediation tools between the transmitter and the receiver. The digital interface, the computer, or other devices that digitally transmit the media allow access. What is essential, in addition to being connected through this access, is the mediation between the transmitter of content and information and the receiver of learning and knowledge construction. This process characterizes the importance of the media mediating the relationship between transmitter and receiver. On the one side, the transmitter—the individual user or a person representing an institution—issues the content, while on the other side, the individual receives and interacts with this content and builds their knowledge. Moreover, the content source is fundamental to the quality of knowledge construction, for which the media are also essential. Media convey messages and content, whereas the technologies enable the development of these media, which are the vehicles, the mediators of the content. In this media configuration, the meaningful participation of designers and architects is supported using technologies that design virtual environments. Some examples of media applied in this online digital interface are animations, which can display 2D or 3D images in virtual, augmented, and mixed realities.[5] However, the same technologies can produce both information and misinformation. To this end, the MIT Center for Advanced Virtuality educates the public on the dangers of deep fakes with the film *In Event of Moon Disaster*, "tackling the misinformation epidemic" (Harrell 2020).[6]

Halsey Burgund of MIT Open Documentary Lab expects this project to lead to an enhanced understanding of how false information is produced in today's media. He stated that "with more understanding and diligence, we can all reduce the likelihood of being unduly influenced by it" (MIT Open Learning 2020). In this sense, it is of the utmost importance to offer relevant knowledge because the distinction between fact and fiction will only become increasingly more challenging in the face of virtual realities. Animation is a resource for presenting content and simulating an abstract situation and architectural spaces using several other media, such as audio, video, text, and images. Images can be distributed in print or digitally, while animations should always be in a digital format. A video expresses a representation that can be of characters and realistically reproduces sequences of images. It can also be used for recording tutorials, especially for procedural content. A game is any activity in which the player's profile—or avatar—and rules can be applied in a restricted or accessible environment. Thus, the information society is followed in a dynamic process through associations, experiences, and the construction and reconstruction of ideas to acquire knowledge in the past and present relationship, and vice versa, with future goals. New technologies have brought digital resources and numerous behavioral and cultural transformations in the forms through which they mediate information (i.e., content). According to the Digital 2020–2021 Global Overview Report,[7] Internet access is already available to billions of Internet users worldwide. These transformations are even related to data privacy concerns. According to the startup Global Web Index (GWI), which provides support for research and case studies on Internet use around the world, a recent survey[8] on behavioral changes regarding the vulnerability of privacy on digital platforms revealed a reduction in concern about the use of surveillance technology, especially in the countries most affected by the COVID-19 pandemic or for security reasons. However, digital technologies also offer protection for user data. GWI's research indicates that Europe has been at the leading edge of such data protection. Even with the increasing number of applications with devices that record and are integrated through Bluetooth, as Figure 9.2 shows, the benefit is more significant due to the notification of individuals at risk. Furthermore, these surveys have revealed that online privacy is evolving and maturing in consideration of the mishandling of user data through new privacy and rights policies and regulations.

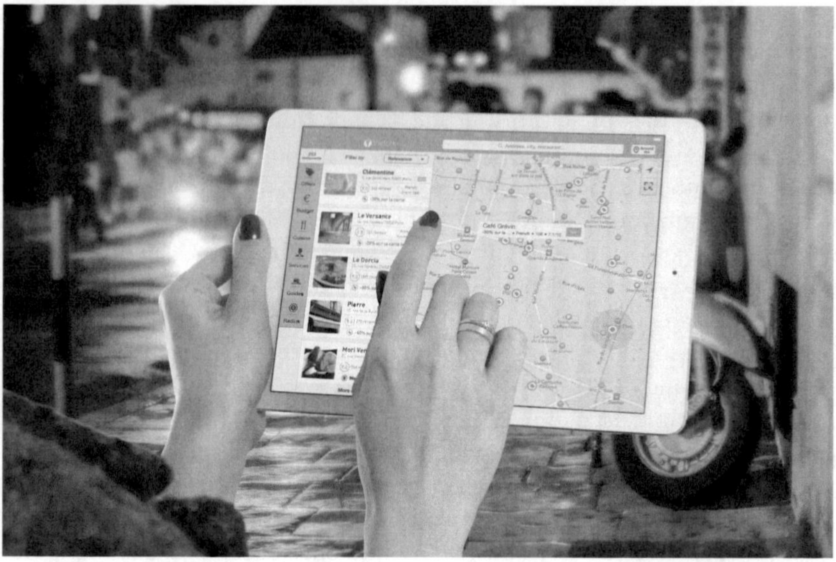

FIGURE 9.2 A city with the availability of public wireless connectivity.

Accordingly, optimistic predictions should be viewed with caution as data tracking methods, although effective, are full of potential privacy issues due to how sensitive data are identified, stored, and processed. On the one hand, this concern has not undermined public demand for these applications, based on the premise that these benefits are essential to public health because they outweigh the privacy risks. On the other hand, the challenge is to overcome advertising and marketing as sponsors of digital resources in all of their ramifications. However, many resources allow these advertisements to be omitted as a form of choice for the user. With these considerations, one realizes that behaviors are being adapted and transformed. Consequently, considering the continuous technological development, cities are becoming information systems. Information is available in real time, as stated by Antoine Picon (2015, 39), who added that "within these systems, relationships between physical infrastructure, service offers and users are being reconfigured, to work toward improved reactivity and greater flexibility of use." In this sense, visualization is essential for the communication process between the different types of simulations by AI concerning the paradigms in their applications, as mentioned earlier. Thus, the moving image in the sense of flow and connectivity is fundamental to the goals of a smart city. As stated by Michael Batty in *The New Science of Cities* (2013), "our science of cities, since resource constraints and different expectations dictate the need for simplicity in design and communication. All this implies fast, simple, visual, and accessible models." Moreover, in the chapter "Urban Simulation" of Batty's work (2013), he further emphasized the importance of visualization in the face of the complexity of cities and their forms of communication, adding the following: "In using the model for integrated assessment, which consists of chaining different models together across different spatial and temporal scales, different kinds of expertise are required." Consequently, the interdependence between social, political, cultural, and economic structures for decision-making involves the dynamics of the city through communication strategies, whereby Batty highlights that "visual media makes it easier to communicate model structures and outcomes to other scientists involved in models that require different disciplinary and professional expertise" (Batty 2013).

Images and texts are developed, transmitted, and received from all areas, thereby conditioning us amid a polysemy of written and visual language. Therefore, knowledge is essential for the discernment of information. With the development of the technological industry in society, the spatial experience, meaning the use of space through technological equipment or even virtual environments, has intensified and helped individuals in their daily lives. However, many people still face difficulties in adapting to the constant technological changes. In this regard, McLuhan had already perceived the effects of technological development in society and understood that resistance to new technologies occurs because the innovation is not familiar. Thus, what is not yet known can generate uncertainties, mistrust, and insecurity and may even represent a threat to more traditional means and resources. In general, it is the point of global society to illustrate the dynamics of how individuals in some way alienated, conditioned or not, also profit from continued investment in digital technology. Their pleasures and behaviors are converted into actions that maintain the system's functioning and continuity. Accordingly, citizens have followed the World Wide Web since its emergence and the whole digital system's codifications and algorithmic rules for the future. Benjamin (1939) understood that reproduction techniques would be meaningless if they cannot increase our reflections or be integrated into our experiences. If these techniques were to replace lived experience, we would lose our senses. According to McLuhan (1964), it is essential for our orientation under the effects of media and new technologies that we do not change our goals due to the newness mentioned earlier. Otherwise, we will only be conditioned to a constant and, perhaps, illusory updating by the latest technologies and resources. Nonetheless, the novelty will always be present. Thus, McLuhan stated that it is not that we must prepare ourselves for the transformations, in the sense that we do not need to be conditioned into a constant updating and monitoring of all of the technological novelties, but rather we must consider what is essential for our evolution. The fundamental point is that new technology merely provides another form for the same content. However, the most significant transformations are cultural ones, which we must understand. This notion is captured in McLuhan's best-known statement—"the medium is the message" (1964). In this way, individuals are transformed in an environment that is not supplemented but modified by constant technological innovations, transcending the time and space between reality and virtuality. In this context, visual fragments configure the images displayed in the public space globally through several configuration techniques, codifications, and optimizations when conveying virtual time and space concerning the physical urban environment. Hence, the considerations concerning the urban image are in motion and connected in all dimensions. The digital world is just an extension of physical urban space.

In Chapter 10, titled "The Arts' Function: Social Inclusion and Sustainability," the focus is on seeking environmental efficiency, considering that the arts achievement also expects solutions. Therefore, Chapter 10 discusses architecture and design with its inherent relationships with art, technology, society attuned to society's needs, and the creation and recreation of new habits and functions.

Notes

1 See Global population, Digital 2022 Global Overview Report, accessed May 3, 2022, https://datareportal.com/reports/digital-2022-global-overview-report.

2 See Global mobile users, Digital 2022 Global Overview Report, accessed May 3, 2022, https://datareportal.com/reports/digital-2022-global-overview-report.

3 Ken Huffman:

> Like many codes, including the one named after Samuel Morse, Huffman's creation tried to find a way to assign the shortest codes to those characters used most, the longest codes being reserved for those used rarely if at all. This process was carried out by forming a so-called coding tree, in which the probability that a number, letter or another character will occur is designated as a leaf on a tree.
>
> "Profile: David A. Huffman: Encoding the 'Neatness' of Ones and Zeroes," *Scientific American* (September 1991): 54–58

4

> The Joint Photographic Experts Group (JPEG) committee has a long tradition in the creation of still image coding standards. JPEG is a joint working group of the International Standardization Organization (ISO) and the International Electrotechnical Commission (IEC). More specifically, the JPEG committee is Working Group 1 (WG1), Coding of Still Pictures, of JTC 1's subcommittee 29 (SC29), Coding of Audio, Picture, Multimedia and Hypermedia Information.
>
> About JPEG, accessed May 17, 2022, https://jpeg.org/about.html

5 Alan B. Craig and William R. Sherman:

> As the required technology has become widely available and of adequate quality, the medium of virtual reality has become a feasible tool for accomplishing research, and not just a topic of research for its own sake. Beginning in the late 1980s and early 1990s, enough technological advancement had taken place that more centers of research (in business and academia) could afford to experiment with VR. Now we have reached the point of technology being inexpensive enough not only for larger research facilities, but for the mass market as well. Thus, using and creating VR content has become largely cost-effective at the consumer level. This book explores what is required to develop virtual reality applications for real-world uses in areas such as science, industry, art, education, and medicine.
>
> *Understanding Virtual Reality Interface, Application, and Design* (Cambridge: Morgan Kaufmann, 2018), https://doi.org/10.1016/C2013-0-18583-2

6 D. Fox Harrell, professor of digital media and of artificial intelligence at MIT and director of the MIT Center for Advanced Virtuality, part of MIT Open Learning, said, "media misinformation is a longstanding phenomenon, but, exacerbated by deepfake

technologies and the ease of disseminating content online, it's become a crucial issue of our time." MIT Open Learning, accessed April 2, 2021, https://news.mit.edu/2020/mit-tackles-misinformation-in-event-of-moon-disaster-0720.

7 See Global Internet users, Digital 2022 Global Overview Report, accessed May 3, 2022, https://datareportal.com/reports/digital-2022-global-overview-report.

8 See Online Privacy, "Consumer Trends Report in 202," accessed May 3, 2022, https://www.gwi.com/reports/trends-2021.

Bibliography

Bardini, Thierry. *Bootstrapping: Douglas Engelbart, Coevolution, and the Origins of Personal Computing*. Stanford, CA: Stanford University Press, 2000.

Batty, Michael. *The New Science of Cities*. Cambridge, MA: MIT Press, 2013.

Baudrillard, Jean, and Jean Nouvel. *The Singular Objects of Architecture*. Trans. by Robert Bononno. Minneapolis: University of Minnesota Press, 2002. Originally published as *Les objets singuliers. Architecture* (Paris: Calmann-Lévy, 2000).

Benjamin, Walter. "Das Kunstwerk im Zeitalter seiner technischen Reproduzierbarkeit" [The Work of Art in the Age of Mechanical Reproduction]. 1939. Reprint, *Medienästhetische Schriften*. Frankfurt am Main: Suhrkamp, 2002.

Craig, Alan B., and William R. Sherman. *Understanding Virtual Reality Interface, Application, and Design*. Cambridge: Morgan Kaufmann, 2018.

Dietterich, Thomas. "Steps Toward Robust Artificial Intelligence." *AI Magazine* 38 (2017): 3–24. https://doi.org/10.1609/aimag.v38i3.2756

Friedman, Avi. *Fundamentals of Sustainable Urban Design*. Cham: Springer, 2021.
https://doi.org/10.1007/978-3-030-60865-1_20

Gonzalez, Rafael C., and Richard Eugene Woods. *Digital Image Processing*. London: Pearson, 2018.

Harrell, D. Fox. MIT Open Learning. 2020. Accessed April 2, 2021. https://news.mit.edu/2020/mit-tackles-misinformation-in-event-of-moon-disaster-0720

Hoppin, Joni, and Leandro Berti. "Como a indústria chegou até as simulações industriais com IA?" [How Industry Came to Industrial Simulations with AI?]. *Aquarela*, January 26, 2021. https://www.aquare.la/como-a-industria-chegou-ate-as-simulacoes-industriais-com-ia/

Huffman, David A. "A Method for the Construction of Minimum-Redundancy Codes." In *Proceedings of the IRE*, vol. 40, no. 9 (September 1952): 1098–1101. doi: 10.1109/JRPROC.1952.273898.

Huffman, Ken. "Profile: David A. Huffman: Encoding the 'Neatness' of Ones and Zeroes." *Scientific American*, September 1991, 54–58.

Knuth, Donald E. *Selected Papers on Design of Algorithms*. Stanford, CA: Stanford University, Center for the Study of Language and Information, 2010.

Koponen, Jarno. "A New Hope: AI for News Media." *Tech Crunch*, 2018. https://tcrn.ch/2NMJRME

Lévy, Pierre. *Les Technologies de l'intelligence. L'avenir de la pensée à l'ère informatique* [*The Technologies of Intelligence. The Future of Thinking in the Computer Age*]. La Découverte: Paris 1990.

McLuhan, Marshall. *Understanding Media: The Extensions of Man*. New York: McGraw-Hill, 1964.

Picon, Antoine. *Smart Cities: A Spatialised Intelligence*. Chichester, West Sussex: Wiley, 2015.

Picon, Antoine, and Thomas Shay Hill. "Is the City Becoming Computable?" In *Architecture and the Smart City*, edited by Sergio M. Figueiredo, Sukanya Krishnamurthy, and Torsten Schroeder, 29–42. London and New York: Routledge, 2020.

Santos, Marcos. *O que é inteligência artificial* (IA) [What Is Artificial Intelligence (AI)]. *Aquarela Advanced Analytics*, May 8, 2018. https://www.aquare.la/o-que-e-inteligencia-artificial/

Schröter, Jens. "Medienästhetik, Simulation und 'Neue Medien'" [Media Aesthetics, Simulation and 'New Media']. *Zeitschrift für Medienwissenschaft* 8 (2013): 88–100. https://doi.org/10.25969/mediarep/702.

Wiener, Norbert. *The Human Use of Human Beings: Cybernetics and Society*. Boston, MA: Houghton Mifflin, 1950.

10

THE ARTS' FUNCTION

Social Inclusion and Sustainability

The analysis and discussion of the function of art concerning social inclusion and sustainability are first motivated by the factors related to the different contexts and cultures for understanding the current moment. Therefore, it is equally crucial to understand the development of civilizations and scientific and technological evolution as processes and successive challenges that depend on sociocultural values, beliefs, and each historical moment. Since the beginning of the 20th century, new art forms and techniques have been essential for cultural transformations related to the rupture of values, the constant search for solutions to social problems, and changes in habits and behaviors. Thus, the most critical social function of art and culture has been to create a balance between human beings and their environment. This balance through art is achieved primarily through the knowledge and understanding of social reality that people acquire, which consequently are accessible through communication and information technologies, representations, and dissemination by the media and through other means. The consequences of these technologies enabling image production resources build, through the media, a collective perception and appropriate forms of individual perception to form an image of the historical and cultural context of the cities and visualizations of the future (Argan 1983; Holtorf and Högberg 2021).

Hence, remembering our origins and evolutionary goals, such transformations are consequences of human achievements in searching for the conquest and domination of nature while occupying spaces, building their habitations, communities, and forms of mutual coexistence. Over time, with discoveries, beliefs, wills, techniques, and inventions, among so many other capabilities, humans developed their intelligence, forming civilizations and cultures. Consequently, this chapter's discussion starts from the following question: What is art for in a cosmopolitan society? This question is related to the ambitions in a market society in many segments, culturally distinct, of large metropolises regarding

DOI: 10.4324/9781003314943-11

artistic and cultural production. Regardless of whether one is an artist, designer, author, actor, or musician, among others, this realization depends on receptivity, which must be understood and interpreted to meet the purpose of its creation. In general, to establish an exchange of values, almost always in search of reward and profit, either by designing solutions or by questioning, everyone finds the same goal, namely, to solve the problems of society. Since the development of the market system, art has ceased to have these same concerns; however, when this occurred is unclear. Aside from ideologies, what prevails is the development of and will to create new techniques, resources, and methods—novel solutions for numerous situations that have not yet been addressed.

To this end, a brief historical notion of the relationship of the arts with the development of cities is presented in the first section, titled "Cities and the Arts." In the second section, titled "The Construction of the Future: Overcoming the Culture of Waste," contemporary challenges aimed at social equality and the reduction of greenhouse gas emissions are addressed with an emphasis on design (Whiteley 1993; Bürdek 2015) as well as the agreements of the United Nations (UN) Climate Change Conference in pursuit of meeting the Sustainable Development Goals (SDGs). In the third and final section, titled "Climate Change, Natural Resources, and the Arts," the focus is on risks to society concerning climate change and the search for an awareness of them through the arts in their critical, imaginative, and revolutionary forms, thus bringing the public closer to planetary reality (Beck [1986] 1993; Welsch 2019).

Cities and the Arts

To talk about the function of art is to think about societies and, mainly, to consider artists, architects, and designers in their realizations and representations of the necessary values for the cultures in their distinct realities. In this regard, reference to the art historian Giulio Carlo Argan (1983), specifically his work *Art History as the History of the City*, situates this approach to the role of arts in its method of exploring the urban space and presenting art in its transforming process.

With origins in the Roman Empire, the arts formed numerous sculptures, frescoes, and architectural and decorative monuments to honor their patrons in public space rather than express the artistic sensibilities of their creators. Mural painting was another feature focused on Greek indoor and outdoor decoration for creating a spatial illusion that extended beyond the mural's surface and the Roman mosaics distributed throughout the Roman Empire. Our ideas about the public and private buildings of ancient Roman architecture were made possible by traces of the writings of that time compiled by Vitruvius at the end of the 1st century b.c.e. Roman theaters and amphitheaters first appeared at the end of the Republican age and were situated on natural slopes. They were built on a structure of pillars and vaults and installed in the heart of cities. Throughout Rome, statues and sculptural reliefs adorned public and private buildings. The arches of triumph, erected in all parts of the empire, stand out as monuments among the most important reality and technological evolution. All world issues are part of the visual configuration of this cultural space concerning the past and the future with historical marks and new visions.

Above all, Greek architecture and sculpture brought the discovery of perspective effects, the contemplation of space, and the use of light. From the 5th century b.c.e. onward, representation developed in art through the constant refinement of form. This development followed a gradual period of corrections guided by the observation of nature as reality until late antiquity. This development of technology use in architecture and sculpture continued with the Romans. As a result, Roman architecture represents the most outstanding contribution to Western traditions in its broadest sense. Nevertheless, it was not until the Renaissance period, when the dome was reconsidered by Filippo Brunelleschi in the architectural practices of Santa Maria del Fiore (1296–1436), inspired by the ancient dome architecture of the Pantheon in Rome, that the dome came to dominate the center of the capital. The hierarchical significance of the relationship between architecture—the dome—and society stands out for its symbolic value and status indicator in the social hierarchy. Sculpture, still as an extension of architecture, still hierarchically represented the statesmen.

From the 20th century onward, sculpture began to seek a nonrepresentational form concerning architecture, which demanded boundaries to distinguish one from the other. However, this was still due to a lengthy discussion about whether works were technical or aesthetic, also considering the canons of the fine arts. Nonetheless, statements about the role of architecture in the early 20th century, as one of the forms that were more spiritual than material in terms of art, whether for the art of building or for decoration or shelter, functionality or allegory has a reference in Le Corbusier, who confirmed that architecture is a fact of art, a phenomenon of emotion, apart from the questions of construction or beyond. The building is for taking hold; the architecture is for emotion.[1] In addition, Le Corbusier recognized this time through painting and sculpture in their new directions for urban life, highlighting that the works of painters and sculptors will be placed within the framework of the times, and that their art will be admitted and understood everywhere.[2] In addition, the image in public space has different forms of representation beyond the arts. However, as an object of study in aesthetics, art is the main object of analysis concerning the configuration of the public image when referring to architectural spaces and monuments that concern the great material, social, or spiritual activities of human life. An example is found in the role of the narrative and persuasive intention through the public image in the story of Trajan's Column in Rome, one of the oldest visual narratives. Situated in Trajan's Forum, this epic work from a.d. 113 illustrates the conquest of Dacia, today a region of Romania. The monument characterizes the intention to immortalize power by influencing image in public space. It is a work with propaganda purposes and was admired by Napoleon. After the battle of Austerlitz, Napoleon also built a column—the Vendome Column in Paris—with the same intentions as Trajan (Wagner 2013, 190).

The relation of architectural works to painting, literature, music, theater, and cinema offers sensitive aspects of the representation of the human experience in its urban environment. These aspects concern aesthetic values not only in the art of building but also in art as a human accomplishment and an object of aesthetics. The meaning would not be in the moment of execution but in the formal result of the work or the visual narratives, by cinema or literature, concerning the urban or natural environments, like each image carved in Trajan's Column (as a cinematographic storyboard). Regardless of which technique is appropriate and compatible with each era, one can observe how visual narratives have been memorized for a long time, stimulating human imagination and creativity. In the development and technical improvement for achievements until today, humanity finds itself amid all available technological resources for producing static or moving images, or images in three or four dimensions, as presented in Chapter 9.

Nonetheless, contemporary urban artists work in situ. They meticulously analyze the place conditions (the scale, user, and complexity of the context) since their work's success depends on the observer's reception. With this, artists have broadened their mediums. They have also begun to build by incorporating new sources of reference, such as science, biology, construction, lighting, decoration, sound, fashion, film, and digital technologies. The transition from temporary installations to permanent buildings establishes approximation with architecture, mainly in what refers to the way of conceiving space. Art and architecture's boundaries have become fuzzier as both are inspired by the physical experience of the subject determined by the nature of the place. Architecture has always been, by definition, public. However, the contextual transformations of the past 20 years have led this discipline to a process of adaptation (just like art). Art in public spaces deals with the relationship between humankind and the world, the subject, and the city, given the urbanistic area facing and affecting such bonds. No matter what is programmed, planned, or designed, the object is always human existence as social existence (Argan 1983). To think about the cities' past and their cultural and artistic legacy today is not to resist the innovations aimed at a technological future in which cities are directed. Instead, one seeks to predict the consequences of this development to build a future. Nevertheless, what is the way to a city, considering the preservation of fundamental human values and their monuments? Would the arts contribute to these social values being respected?

Recently, Holtorf and Högberg (2021) highlighted the future of cities concerning the past and how investments in cultural heritage preservation are conducted. As far as historical fact is concerned, it implies the past and the construction of the future. The contemporary concept of the future did not emerge until the early modern period, according to Lucian Hölscher (2016) in *The Discovery of the Future*. Therefore, historical, social, and political issues attribute values and meanings to cultural heritage. According to the professor and president of the research commission at the Museum of Contemporary Art, University of São Paulo (MAC USP), Edson Leite (2011), integrating cultural heritage as an expression of social development in the constitution of collective identity influences the practices linked to its protection. Consequently, thinking about the future, the memory preserved by the conservation of material goods and by the registration of immaterial goods can gain greater meaning if conveniently inserted into social and collective life by digital technologies in several virtual platforms.

The Construction of the Future: Overcoming the Culture of Waste

In principle, all artistic expectations for meeting the global needs of social inclusion, democratization, and sustainable and green economies depend not only on their projects but also mostly on good political will and new cultural and consumption habits. Moreover, arts achievement also expects solutions with social projects. Such projects concern cultural differences, education, and the fight against poverty and hunger as the fundamental basis for broad and dynamic innovation in today's "culture of waste" system of our consumer society. One of the most significant works critical of society is precisely titled *La société de consommation* (*The Consumer Society*) by Jean Baudrillard (1970). This book, originally published in Paris, featured an artwork by Daniel Spoerri on the cover of its most recent edition (2005). The work is an assemblage titled *Phallen, Fallen-Bild* (1989), which depicts the end of a large meal with debris scattered on a plate and over a table. From this image, even before the introductory text and in-depth reading, one can deduce the main problem of this consumerist system and those living in abundance—namely waste.

Therefore, considering theory and practice, the goal of this analysis is to understand the interdependence between architecture, art, and design in its technical essence. In this sense, the interdependence is a process through which it is possible to understand the influence of artistic creations on commercial products. Of course, the most considerable significance is the direction that designs have taken concerning the post–World War demands of globalization and the Fordist economy model's relationship with the current condition. It is worth considering the post-war Fordist world economy versus its coincidence with the beginning of a long process that led to the end of the Fordist system itself and its replacement by an economic regime that David Harvey called "flexible accumulation," which we know as late capitalism. In Harvey's work *The Condition of Postmodernity: An Enquiry into Origins of Cultural Change* (1990), our present system differs from the Fordist system through the decentralization of capital. In the Fordist order, power and wealth were measured by the country's productive capacity, company, or individual, whereas in this flexible system, the consumer's power of consumption is gauged.

Further, formal innovations no longer follow function in their unique and determining purpose but are still associated with consumer identity values, meeting their needs and lifestyle, within criteria aimed at the sustainable development of diversified and fragmented production. There is no longer a Fordist sense of mass output without concern for society's cultural and specific differences. Cultural stratification and consumption on a global scale, as the classification between developed, developing, and underdeveloped countries, are no longer the usual social, political, or economic perspective. Even the denomination of "first world," "second world," and "third world" has not made sense in 30 years, since 1989 and the end of the USSR and the communist system, which were understood as "the second world," while the so-called "first world" were NATO countries belonging to the Warsaw Pact. The other countries were the third world. Today, economists and sociologists orient their analyses based on center and periphery relations, enabling a comparative dynamic where their parts interact, diversify, and restructure fragmented and distinct values to maintain the global economy in its inherent differences and similarities in every culture. Currently, the world is again experiencing a crisis between global powers in the face of the relations between Russia, Europe, and NATO, highlighting a possible new world order. This context primarily implies a technological acceleration of renewable energies due to Russia's gas independence goals. Nevertheless, we already know that since the 1970s, the oil crisis and environmental issues concerning the scarcity of natural resources are the main problems. The oil crisis has been the leading cause of an awareness of the industries that concern the world economic system, as it led people to consider energy scarcity a significant threat. Therefore, research and projects focused on new renewable energy sources and organizations aim for new conditions to preserve the environment, current resources, and socioeconomic conditions to ensure future generations. Among the leading organizations are Friends of the Earth, founded in 1969, and Greenpeace, founded in 1972 to ensure future generations' access to a healthy environment.

However, new technologies give design and industry the chance to realize the ideals proclaimed by environmental movements. This functionality emerged with the Bauhaus school of arts in Germany in 1919, and later with globalized production, from the international style in the United States and the Ulm School of Design in Germany (Hochschule für Gestaltung Ulm) in the 1950s. The role of innovation has been central to dealing with the individual's aspirations in society and the demand of a standard and mass industry. However, the market development situation after the Second World War took the role of design to the ultimate consequences for the status of the obsolete object today, favoring a market for goods and consumption. Thus, design can offer the chance to recycle its configuration norms, deconstruct its values and reconfigure them creatively, and combat problems related to waste, pollution, global warming, and social inequalities. On the other hand, this creative practice is related to the leading social criticisms of the industrial and consumption context and its consequences. Since the Stockholm Conference in 1972, the UN has sought to make society aware of the relationship between human beings and nature. Many solutions aimed at using renewable energy and reducing greenhouse gas emissions have been presented, through targets set in a series of agreements aimed at sustainable development. The significant challenges are the logic of the global market and the call for innovation for solutions to preserve nature, especially concerning the material aspects of society—technological and economic development. Above all, social inclusion and equality must be considered primordial factors. In this sense, design in its function between art, technology, and society, in applied social sciences, produces knowledge and aims for solutions to social demands with projects based on economic decisions.

Moreover, some sociologists hold—to some extent within consumer society, despite its different social classes and economic inequality—that the scope of a genuinely democratic consumption does not seem possible. Therefore, some design and society experts understand that social inclusion is favored within this market logic. Nigel Whiteley (1993) analyzed the effects that abundance and marketing have on the design and movement of green consumption. In his studies, without contradicting the design style, he has highlighted design as solutions considering the "social utility" and environmental effects of a product. He understands that consumer-oriented design, stimulating consumer society, offers the democratization of private property and luxury compared with scarcity and subsistence-based nations. According to Whiteley, if discarded still in good condition, a new or used product can still be consumed as "second-hand." In this sense, for Whiteley, the importance of ownership is given, regardless of the current or outdated state of a product, and he believes that social inclusion exists. It should be noted that this consumption does not coincide between the bourgeoisie and the economically subordinate classes. Despite the technological speed at which products become outdated, the subordinate classes almost always have the alternative of consuming "the trash of luxury"—that is, things that the bourgeoisie no longer wants and even discards due to their technological obsolescence, physical form, and functionality. Along these lines, in the current international contemporary art exhibition, Documenta 15, the group The Nest Collective presents the installation *Return to Sender—Delivery Details*. This installation was built with used clothes in the public space of Kassel, Germany, where the event takes place, intending to confront the current context of the global consumer society visually. According to The Nest Collective, *Return to Sender—Delivery Details* "echoes the true breadth of the uncontrolled consumption in the Global North, and the stringent regulations that overtly force the movement of these materials to the Global South"; consequently, this artistic group manifests visually through their art "the business and headache of less industrialized nations to deal with, as both waste and non-waste."[3]

Moreover, in emerging economies, creating a more inclusive society concerning the green economy and social and environmental sustainability remains an enormous challenge. Outdated and environmentally harmful technologies are still the primary means of production that provide economic stability and employment in growing economies. The main factors in the evolution of science and technology are industrial modernization and design, which provide innovative solutions. However, it should be noted that designers are not the inventors of these technologies; rather, they design creatively through them. Designers are the interpreters of new technologies, and they realize social values (Bürdek 2015). In general, architecture—design with its inherent relationship with art, technology, and society—should be attuned to society's needs. Therefore, sustainable development is the type of development capable of meeting the needs of the current generation without compromising the ability of future generations to meet their own. The green economy proposes that, in addition to productive and social technologies, other means should be created through which essential factors linked to socioenvironmental sustainability, still ignored in economic decisions, can be addressed. For example, the "inclusive green economy" includes programs that promote environmental preservation or recovery, supporting segments of the population earning an income from recycling solid waste. In this sense, the goal is to consume less and more consciously with initiatives aimed at rationalizing and optimizing products for low energy costs or renewable energy and the use of renewable and environmentally friendly materials. Thus, technological and scientific development in partnership with industry has intensified research into alternative energy solutions. Independence from nonrenewable energy sources is the new economic direction of many countries. However, some countries still resist this demand based on political and economic interests. The UN strongly supports the most important treaties that condition decision-makers to follow sustainable development and a green economy. Since the Stockholm Conference in 1972, the UN has sought to raise society's awareness of the relationship between humans and nature. Many solutions for the use of renewable energy and the reduction of greenhouse gas emissions have been presented in the form of targets set in a series of agreements—the current one being the 2015 Paris Agreement adopted at the COP21 in Paris—aimed at sustainable development.

Thus, the design must make the following items feasible, albeit for their production management qualities: A calculation of the product's life cycle; certification of a fair relationship of trade and ecological standards from the origin of the raw material to the consumption of the final product; an aim to guarantee a return to the production chain, without waste or polluting residues; and the assurance of sustainable development. Design counts on its allies—marketing and advertising—if there is political will for well-intentioned decisions through the agreements between national and international markets. However, one question is indispensable, especially since thinking about innovation and consumption are also necessary: Can technological advances offer truly democratic consumption? Democratizing access to the most sophisticated, expensive technologies is in the interests of everyone. However, only one idea of the democratization of consumption is apparent; that is, both the more affluent social classes and their respective strata as well as other more modest social segments can consume the same products in different ways and, in some cases, with the same design, in their aesthetic and functional aspects.

Understanding the reasons for this is easy. A product that is technologically outdated and with, among other things, an established design but now unappreciated by the elite is not an imitation of anything. The macroeconomic theory that the increase in industrial production, allied to the participation of financial capital, "brings social classes closer" and "democratizes" consumption should continuously be reevaluated. Historically, the strategies that guaranteed the strengthening of the North American economy during Second World War, based only on the concept of simple stylistic obsolescence, no longer ensured the continuity of the system as it was also necessary to offer purchasing power to the consumer. The solution was consumer credit in the post-war period with a new economic initiative. In 1950, the credit card appeared. This was the reason for the unprecedented growth of the economy based on a consumerist society that propagated an American way of life, in which abundance and waste are the motivating bases for prosperity. Nevertheless, in the early 1970s, environmental issues related to the scarcity of natural resources became the new reality for a global society, as previously mentioned. Moreover, one must consider that in some way, despite the distinct social classes and economic inequality within consumer societies, the achievement of a genuinely democratic consumption seems impossible. However, some design specialists understand that social inclusion is possible within this market logic, as Whiteley (1993) considered. Although the elite and subordinate classes are part of the same consumer society, some significant differences should be understood. The former live in a world where all novelties appear to be bought and used immediately, whereas the latter live the desire for this acquisition primarily stimulated by the advertising image. Then, the desire is fulfilled, where possible, by the purchase, almost before the aesthetic and technological obsolescence of the product—that is, the design that has lost its charm and no longer seduces consumers with greater purchasing power. Consequently, the main factors to consider are the speed at which new technologies emerge and their cost and benefit. With this, one must think that the consequences of this dynamic relationship between innovation and the market still have their bases in what the American socialist and economist Leo Huberman affirmed in his work *Man's Worldly Goods: The Story of the Wealth of Nations* (1936) when he stated that all and any work should generate wealth for the State and society; otherwise, social problems and economic crises would result. We are now in 2022, almost a century after Huberman's statement, going through various crises and social problems to see what he prescribed in theory. Thus, in their evolution, science and technology are the main factors for industrial modernization, and design and architecture are, in their practice, the creative means of solutions for public infrastructure and all objects and spaces that are meaningful and useful to society. In addition, the arts, in general, provide new forms of worldviews.

Climate Change, Natural Resources, and the Arts

Among the most significant factors concerning climate change problems in social awareness is the risk factor. However, this factor is not well understood. According to Ulrich Beck ([1986] 1993), a solution to inequality and global justice as good political will is essential for a global solution. Thus, one can consider

Beck's observations regarding the discussion of climate change as a priority on the political agenda in a global society at risk. These observations guide the reality of technological development and science as possibilities. However, the force of this discourse on reality has particularly convinced the market, converging in the perennial development of the "green industry," supported by social and environmental movements.

The UN Climate Change Council recently took another step, with the Glasgow Climate Pact in November 2021, toward fulfilling the political agreements already established in the Paris Agreement of December 2015. However, this step needs to be updated following global development. Immediately critical is the need to start considering the emerging problems of a "green image" in the search for the technical conditions of the present through new technological means of production, new consumption habits, and projects aimed at innovation—including the issue of climate change relating nature in contrast to urban life.

Thus, Beck suggested that the global solution is found by using not only high-level technology but also new forms of participation and contribution, making society's risk factors common to all cultures. To Beck, risks are not synonymous with catastrophe but rather anticipation. They exist in a permanent state of virtuality and become topical when anticipated through various visualization techniques, especially those used by new media and the arts. Therefore, environmental issues in the arts began to emerge with the environmental and ecological movements in the 1960s and 1970s.

Land art is one of the art forms that stands out as an artistic and aesthetic movement regarding the relationship between humans and nature. The first works appeared in the United States with environmental and ecological activities that sought to offer an aesthetic experience involving the observer in the nature of the art. Robert Smithson's *Spiral Jetty* (1970) is the most significant example, which he built with earth and stones on Utah Lake. Land art has nature itself as its material support. Regardless of this aesthetic experience with the dimension of the sculptural work, the most critical element is evidence of human intervention in nature. In this sense, the present author's observation highlights the concept of the Anthropocene through artistic intervention. After more than half a century, this aesthetic experience is made possible not by immersion in the work in nature but by forms of visual transfer through photographic and video images. This sense was highlighted by Christiane Heibach (2016) in her article "Von der Land Art zur Klimakapsel: Ökologische Utopien in der Medienkunst" (From land art to climate capsule: Ecological utopias in media art), which concerned works that intentionally create a field of tension between the creative will of the artist and the creative power of nature.

A strong trend of ecology in aesthetic and art theories as well as in artworks is related to the environmental context, especially given the popularity of this topic. In this context, aesthetic and art theories present concerns about the production aspects of the market and the individual values inherent in the human species concerning its nature. Crucial here is the affirmation of artistic achievement with freedom of expression. This freedom offers the transparency of contemporary society and identifies the reality toward which we must move to find the right direction for humanity. According to this scenario, examples of the art that has been created are numerous—we are surrounded by it every day. However, some of the most notable highlights from the international art scene were created in the 1970s and 1980s. In 1982 in Kassel, Germany, during the international art exhibition Documenta 7, Joseph Beuys planted 7,000 trees. This was an artistic action with a vision of climate policy. Before this artistic action, another work of Beuys regarding the relationship of humans with the environment, industry, technological development, and territorial domination was his performance *I like America, and America likes me* (1974). The performance, held at the René Block Gallery in New York, involved Beuys staying in a room for three days with a coyote. The coyote represented the ancient spiritual energies of the Native Americans and their extermination by the colonizing civilization that dominated the North American continent. The performance offered a socio-aesthetic critique that drew attention and historical awareness to socioterritorial conflicts concerning indigenous peoples and former European settlers, culminating in everyday mass culture. However, the performance was not attended by an audience and was instead recorded in photographs by Caroline Tisdall (1976) as well as some video clips. *I like America, and America likes me* is representative of the comparative aspects of human nature (Beuys) and its animal nature (coyote) for dichotomies of interpretation on reason and instinct, artificial and natural, and development and native peoples—in short for the power relations between humans and nature.

Additionally, one of the first artistic interventions in urban space using new technologies, art, and architecture involved building the Central Headquarters of the Affiliated Gas Companies Network, which the American artist James Turrell did in Germany in 1977. Turrell aesthetically materialized nature as an event by using an "intelligent" system to automatically capture climate and thermal factors, activating the building's lighting system. The glass architecture was illuminated entirely without the artist's interference since the external temperature determined the results of the work, just like the Danish artist Olafur Eliasson, who reflects on the city—nature relationship. Eliasson, with his interventions in the Green River series, produced between 1999 and 2001 in several cities around the world, acted directly in the reality of the city. His intervention consisted of adding green pigment to rivers in the cities. Occurring in the domains of the city, his Green River intervention dealt with the fears and doubts of the collective.

In April 2020, Eliasson used virtual reality to present a series of images titled *Earth Perspectives* to honor the 50th anniversary of Earth Day. This action symbolized the viewer projecting a new world. London's Serpentine Galleries commissioned the piece as part of their Back to Earth program. During the COVID-19 pandemic, Eliasson launched a second digital project titled *Wunderkammer.* The true effects of actions through arts and new media, considering social changes, are linked to the collective imagination and the meaning of art throughout its history. However, it is evident how the artists present to the public the emerging issue of what climate change is today, the appreciation of planet Earth, and the preservation of its natural resources for humanity.

Therefore, the images, in their effects resulting from the attitudes of the artists, who add their history to their art, are above all the reality of today in their impact on society. Therefore, what predominates are the probable and actual effects through new means, with an awareness of culture regarding climate change and the threat of a global catastrophe.

Another highlight is the work of the Polish and naturalized Brazilian artist Frans Krajcberg (1921–2017). His works can lead to a nostalgic return to nature through valuing biodiversity. Krajcberg, due to his own history, had a different perception. He stressed the importance of life and human nature and explained that nature gave him strength and the pleasure to feel, think, work, and survive. His works resulted from a direct experience with nature and present the destruction caused by humans. His sculptures demonstrate a still life of deforestation in the Amazon and the Pantanal (Mato Grosso). The artist expressed the value and power of nature above the immediate and commercial values that lead human nature to its destruction, with the mark of the sublime beauty of his sculptures. A few years ago, Krajcberg presented a manifesto. In it, he drew attention to the art market and called on the art world to rediscover a sense of nature, harmony, and reason for an avant-garde creation that values freedom, dignity, and respect. *Le Nouveau Manifeste du Naturalisme Intégral* (The New Manifesto of Integral Naturalism 2013) by Krajcberg and Claude Mollard and *Manifeste du Rio Negro* (The Rio Negro Manifesto [1978] 2003) by Pierre Restany were presented during COP 21 in December 2015 in Paris. *Le Nouveau Manifeste du Naturalisme Intégral* includes, in its many paragraphs, a commitment against the process of the destruction of the planet by humans, which would lead to their own destruction if allowed to continue. Krajcberg did not make his art a product of the market; rather, through his revolt he called for the Amazon to be protected and created to denounce the dangers that exist on the planet. His art aims to raise awareness through the transformation of the human management of nature, and it is art committed to ecological issues to fight to prevent a possible global catastrophe.

Other critical issues concerning arts and climate change have also been addressed in social and artistic contexts. In a recent article by Wolfgang Welsh (2019), titled "Nach dem Ende des Anthropozäns" (After the End of the Anthropocene), the author addressed the position of art concerning the Anthropocene in three ways. First, Welsh related art to the environment, interventions, or the character of a documentary report without offering a solution, only a good intention, emphasizing an already established situation. Second, Welsh focused on human beings and the conditions of their existence on Earth, precisely because of their activities, and also on forms independent of any earthly state, focusing on artificial intelligence (AI) and digital technologies as a possibility for the continuity of human existence. He highlighted expectations of solutions by the arts that represent the technological problems of the Anthropocene. However, one must consider the reality of these artworks, not the reality external to them. For Welsh, this is of course inherently perverse, and he added that severe doubts exist as to whether these technological fantasies are anything more than daydreams. Third, Welsh continued to address the situation of art concerning the Anthropocene (in the sense of logic appropriate to the context of art manifestation) by presenting his aesthetic reasoning about a third form of art related to the end of the Anthropocene—that of a planet that is not dependent on human beings. Welsh exemplified his aesthetic reflection with a series of photos by artist Helmut Wimmer titled *The Last Day* at an exhibition at the Kunsthistorisches Museum in Vienna (2018). Wimmer's photos represent the causes of human exploitation and the destruction of nature due to human culture, which does not respect nature's signs, thus representing a nature that no longer cares about humans demonstrating their dominance.

The opposition of nature and culture involves interactions between the human and nonhuman. In this regard, two aspects are critical in cultural transformation: Technique, as the notion of art, and human beings. Therefore, many possibilities of individual interactivity have been generally realized in their own sociocultural environments through the techniques used in artworks. In such processes, realization always looks for possible creativity linked to the diversity of art configurations. The continuum of nonhuman and human aesthetics remains to be observed in Darwinian fundamentals. Therefore, Welsch (2004) suggested an essential "human aesthetics against sociobiological reductionism." He demonstrated the progression from beauty as a physiological effect without aesthetic function to an aesthetic sense. While the context of the evolution of species is the struggle for life, the evolution of our brain as the source of an aesthetic sense in human culture, in terms of the differences between nonhuman and human pleasure, is beyond the perception of the beautiful. Therefore, the results in both animal and prehuman or human environments enable innovation regarding the acquisition of management skills. Moreover, an underlying support for environmental management and control through human evolution has been the feeling of freedom. Such an innovation as the feeling of freedom is a pleasurable and beautiful thing and came about in response to all of life's struggles. Moreover, in terms of this condition in animals, such a transference of beauty from body to artifacts is a medium for overcoming their natural limitations. Accordingly, Welsch analyzed Freud's theory of sublimation as "the one great motor of human cultural development." Thus, we can understand the origin of human aesthetics in our own condition as animal species. We are concerned with the invention of culture, that is, building artificial environments that matter to nonhuman entities with social effects. In such a way, aesthetic practices can be realized by such arts as painting, sculpture, architecture, design, film, literature, and music. There have been similarities in the attitudes of human beings throughout their evolution regarding the creation of objects that reflect the ideas of mobility and self-extension.

Invention has also been added to the complex subject of the relationship between humans and nonhumans, where both emotion and reason influence the ideal of innovation and are associated with the most attractive part in the art of beauty and its reverse. It is useful to think of the traditional forms in which the concept of the nonhuman manifests itself. For example, in natural representations, those in artworks in the generally accepted style, varying over time and by aesthetic experience; in relation to contemporary art, which is integrated with social critiques; as humanity's resistance to market influences; and as support for a sustainable environment in coexistence with technologies that preserve tradition and nature while searching for innovation. The individual is able to create new realities, establishing the emergence of new forms to represent ideals through digital technology, the Internet, media convergence, and hypermedia. This involves a sense of purpose and the contemporary relevance of new theories in aesthetics, particularly in terms of helping to conceptualize the status of the contemporary image and the daily life of a metropolis. Hence, artistic, cultural, and political interdependence must be understood globally as a significant process for contemporary urban life in the articulations, "acknowledging that climate change is having increasing impacts on people especially in the developing world" at the UN Climate Change Conference in Glasgow (COP26). To this end, social responsibility is one of the primary characteristics of this interdependence in its functional, rational, and objective aspects, not to the detriment of individual sensitive values, their desires and particularities sublimated in consumption, but in the synthesis of the relationship between reason and sensitivity to meet, in all its aspects, the aspirations idealized by society. Therefore, in Chapter 11, titled "Designing for Sustainability," design and architecture practices based on the UN Environment Program, aimed at building awareness to guide decision-makers in following sustainable development worldwide, are analyzed and discussed in terms of the objectives of a smart city.

Notes

1 Le Corbusier: "*L'architecture est un fait d'art, un phénomène d'émotion, en dehors des questions de construction, ou delà. La construction c'est pour faire tenir; l'architecture, c'est pour émouvoir.*" *Vers une architecture* (Paris: L'Esprit Nouveau, Les Éditions G. Crès et C, 1925), 9.

2 Le Corbusier: "Messieurs les peintres et les sculpteurs, champions de l'art d'aujourd'hui, . . . Vos œuvres viendront alors se placer dans le cadre de l'époque et partout vous serez admis et compris." *Vers une architecture*, 9.

3 The news and images of this installation can be seen on The Nest Collective's official website: "The Nest at Documenta 15: Return to Sender," last modified June 14, 2022, https://www.thisisthenest.com/news/2022/6/14/the-nest-at-documenta-15-return-to-sender.

Bibliography

Adorno, Theodor W. *Philosophy of Modern Music.* Trans. by Anne G. Mitchell and Wesley V. Blomster. New York: Seabury Press, 1973.

Argan, Giulio Carlo. *Storia dell'arte come storia della città* [Art History as the History of the City]. Rome: Editori riuniti, 1983.

Baudrillard, Jean. The Consumer Society: Myths and Structures. London: Sage, 1998. Originally published as *La société de consommation: ses mythes, ses structures* (Paris: Denoël, 1970).

Beck, Ulrich. *Risk Society: Towards a New Modernity*. 1986. London: Sage, 1993.

Bürdek, Bernhard E. *Design: History, Theory and Practice of Product Design*. Basel: Birkhäuser, 2015.

Deleuze, Gilles. Cinema I: The Movement-Image. Trans. by Hugh Tomlinson and Barbara Habberjam. Minneapolis: University of Minnesota Press, 1986. Originally published as Cinéma 1: *L'image-mouvement* (Paris: Minuit, 1983).

Deleuze, Gilles. *Cinema II: The Time-Image*. Trans. by Hugh Tomlinson and Barbara Habberjam. Minneapolis: University of Minnesota Press, 1989. Originally published as *Cinéma 2: L'image-temps* (Paris: Minuit, 1985).

Gregori, Thomas R. de. "The Back-to-Nature Movement: Alternative Technologies and the Inversion of Reality." *Technology and Culture* 23, no. 2 (1982): 214–217. https://doi.org/10.2307/3104134.

Harvey, David. *The Condition of Postmodernity: An Enquiry into the Origins of Cultural Change*. Cambridge: Blackwell, 1990.

Heibach, Christiane. "Von der Land Art zur Klimakapsel: Ökologische Utopien in der Medienkunst." *Künste Medien Ästhetik* v. 1, no. 3 (2016): 1–10. https://edoc.hu-berlin.de/bitstream/handle/18452/8143/heibach.pdf

Herbert, Isolde Karen. "Nature and Art: Morris's Conception of Progress." *Journal of William Morris Society* v. 10, no. 1 (Autumn 1992): 4–9.

Hölscher, Lucian. *Die Entdeckung der Zukunft* [The Discovery of the Future]. Göttingen: Wallstein, 2016.

Holtorf, Cornelius, and Anders Högberg, eds. *Cultural Heritage and the Future*. London and New York: Routledge, 2021.

Huberman, Leo. *Man's Worldly Goods: The Story of the Wealth of Nations*. New York: Monthly Review Press, 1936.

Krajcberg, Frans, and Claude Mollard. *Nouveau manifeste du Naturalisme integral* [The New Manifesto of Integral Naturalism]. Grenoble: Critères éditions, 2013.

Le Corbusier. *Vers une architecture*. Paris: L'Esprit Nouveau, Les Éditions G. Crès et C, 1925.

Leite, Edson. *Turismo cultural e patrimônio imaterial no Brasil* [Cultural Tourism and Intangible Heritage in Brazil]. São Paulo: INTERCOM, 2011.

Restany, Pierre. *Manifeste du Rio Negro: Du naturalisme intégral* [Rio Negro Manifesto: On Integral Naturalism] 1978. *Revue Inter, Les Éditions Intervention*, v.1 no. 85 (2003): 18.

The Nest Collective. www.thisisthenest.com

Tisdall, Caroline. *Joseph Beuys: Coyote*. New York: Thames and Hudson, 1976.

United Nations Climate Change. *Glasgow Climate Change Conference*. Accessed April 22, 2022. https://unfccc.int/process-and-meetings/conferences/glasgow-climate-change-conference

United Nations Climate Change. *Renewables in Cities 2021 Global Status Report*. Accessed April 22, 2022. https://www.un.org/en/climatechange/reports

United Nations Climate Change. *The Paris Agreement*. Accessed April 22, 2022. https://www.un.org/en/climatechange/paris-agreement

Wagner, Christiane. "Esthétique: l'image contemporaine: l'analyse du concept de l'innovation" [Aesthetics: Contemporary Image. Analysis of the Concept Innovation]. PhD diss., Universtité Paris 1 Panthéon-Sorbonne, 2013.

Wagner, Christiane. "Aesthetic Experience: Visual Culture as the Masterpiece of Nonhumanity." *On_Culture: The Open Journal for the Study of Culture* 2, Justus-Liebig-Universität Gießen (November 2016): 1–12. http://geb.uni-giessen.de/geb/volltexte/2016/12358/

Wagner, Christiane. "Artistic, Cultural, and Political Interdependence: Cities in Motion Facing the Challenges of Technological Advances." In *Cities in a Changing World: Questions of Culture, Climate and Design*, edited by J. Montgomery. *AMPS Proceedings Series* 24, no. 1, (2022a): 133–142. The City University of New York, New York City College of Technology. https://amps-research.com/proceedings/

Wagner, Christiane. "Aesthetic Appreciation: Natural, Artistic, and Media Effects." *Art Style, Art & Culture International Magazine* v. 9, no. 9 (March 2022b): 129–147. https://doi.org/10.5281/zenodo.6757185

Welsch, Wolfgang. "Animal Aesthetics." *Contemporary Aesthetics* v. 2, no. 1 (2004). https://digitalcommons.risd.edu/liberalarts_contempaesthetics/vol2/iss1/15/

Welsch, Wolfgang. "Art Addressing the Anthropocene." *Serbian Architectural Journal* v. 11, no. 2 (2019): 341–358. https://doi.org/10.5937/SAJ1902341W

Whiteley, Nigel. *Design for Society*. London: Reaktion Books, 1993.

11

DESIGNING FOR SUSTAINABILITY

Design challenges are related to the whole social structure and depend on advances in science and technology. They involve the complexity of all knowledge in practice and theory. The technical ability for intellectual and empirical activities is essential to the design ability for constructing knowledge and materializing the ideas that result in all urban structures today. Therefore, design[1] is discussed as a technical and creative process that is necessary for planning sustainable development. The historical and theoretical approach to design is based on the relationship between art, technique, and architecture in its industrial conception and the sociocultural consequences of its practice in urbanization and technological development. However, to better understand this relationship, focusing mainly on the present day, this chapter discusses design in its technical significance applied in a global dimension in two parts. In the first part, titled "From Origins to Future Objectives," the focus is on the fundamental notions of design related to social objectives and the first conceptions for the construction of the corporate image as a market strategy. The second part, titled "The Role of the Designer," focuses on the challenges faced by the designer in meeting sustainable development. Thus, the creative and realization process is analyzed concerning technological and scientific development. Consequently, this chapter seeks to better understand the creative process, that is, forms, content, and structures concerning the global and local context in its potential for innovation and social transformation. On the one hand, the significant aspects of creative practice in consideration of the sociocultural resources and stimuli from the 20th to the 21st century are contextualized (Bürdek 2015, 2019; Papanek [1971] 2019). On the other hand, this innovative practice is related to the leading social challenges of the current political, economic, and sociocultural context (Bourdieu 1986; Dewey [1916] 1997; Elkington 1997; Beckman and Barry 2007; Norman 2013; Stokholm 2014; Vogtländer and Wever 2015). Finally, design is related

DOI: 10.4324/9781003314943-12

to its objectives and the material aspects of society, that is, regarding the economic, scientific, and technological development in the local and international marketplace.

From Origins to Future Objectives

The Bauhaus school formally existed from 1919 to 1933 in Germany. In its origins, all events indicated that the school would be revolutionary and innovative. Today, however, Bauhaus as an educational institution in the arts field attracts conflicting opinions regarding the work done during its existence. Thus, it would not be possible to reach a unanimous consensus, precisely due to the school's history—brief but significant, complex, and polysemic, with a legacy characterized by meandering. In this sense, the school contributed significantly to the history of design and architecture, including projects that would come later, especially in their participants' dedication to accomplishing novelty in the face of social demands. The Bauhaus focused not only on new products but also on new ways of living. It was a social phenomenon, not a design or an art phenomenon. From 1919 to 1920, from the culture of housing and the social principle of consolidating art to the people, the Bauhaus school influenced society's way of life. However, its influence was focused on meeting the population's interests with products of quality and functionality. Conscious of social needs, this development of functionality was guided in its practice by architects, artists, and theoreticians, who knew people's daily needs. The beginning of a designer's training at the Bauhaus relied on the teachings of Peter Behrens, which formed the introductory course at the Bauhaus (*Vorkurs*). Behrens' aim was to return to the fundamental intellectual principles of all form-making design, allowing such principles to be rooted in the artistically spontaneous aspects and their internal laws of perception rather than directly in the mechanical aspects of a design project. Students would draw and paint natural forms in different media and then conduct analytical studies to explore linear movement, design, and geometric structure. Two of Behrens' apprentices were Walter Gropius and Ludwig Mies van der Rohe. Even Mies van der Rohe's maxim "less is more"[2] has its origin in Behrens' teachings, according to Detlef Mertins' monograph (2014) on some of the most significant works in architecture and design in that context. In addition, before the Bauhaus, the Deutsche Werkbund (German Association of Craftsmen) was founded in 1907 in Munich, which advocated the fusion of art and technology.

Behrens played an essential role in this organization, the first created to infuse manufactured products and architecture with high-quality design. The group's leaders, including Hermann Muthesius, Henry van de Velde, and Behrens, had been influenced by William Morris and the English Arts and Crafts movement, albeit with significant differences. While Morris was repulsed by the products of the machine age, not exactly in a back-to-nature[3] sense but rather as an advocate for a return to medieval craftsmanship to protest the Industrial Revolution,[4] the Werkbund recognized the value of machines and advocated a design to give form and meaning to all things made by machines, including buildings. These designers proposed a social philosophy, a culture encompassing the coexistence of humans and machines (*Gesamkultur*), that is, a new universal culture in a reformed artificial environment. Design was seen as the engine that could drive society forward to achieve this *Gesamkultur.* With successive social and cultural transformations due to technological and scientific innovations, the legacy of these institutions and movements undoubtedly still has an influence. In 1928, Hannes Meyer was one of the leading scholars to advocate for the social engagement of architects and designers, demonstrating that the critical function of design was configuration (*Gestaltung*) with balance, technique, materials, and industrial production conditions that meet social needs. Thus, in short, it was essential to condition social projects to the needs of the population. Therefore, over time and for a politically ideological issue, two fronts were marked by the design culture after the Second World War—one focused on Communism and the other on Capitalism.

In the 1960s and 1970s, social themes influenced the design of the market economy—as did the Russians, Socialist Germans, and Bauhaus school—in the face of the influences of the political-ideological movements of that period. In this case, the development of industry, whether in capitalist or communist society, was a crucial aspect. Unlike other artistic manifestations, the "art" of the designer began to have a restricted relationship between object and user, an interaction, or a dialogue between the two. This "reciprocal appropriation" between designer and objectivity is what differentiates the work of creation. However, what is referred to as "reciprocal appropriation" is only one of the distinguishing features inherent to the process of creation. Apart from this aspect, other segments of design have multiple productive activities. Here, one should register the product design, fashion, and graphics, besides, of course, the work of the architect and the artist, precisely because of the close relationship that has always existed and continues to exist between these activities. They have always aimed at innovation.

The project development process, the design, and the configuration of the relationship between the object and the product, environment, and user are the primary means of technological innovation. It is in this way that we can remember some of the accomplishments of the members of the Bauhaus and, afterward, by other designers. Good quotes exist on exploration methodologies for design projects, relating more to aesthetic and technical design. Here, Bernhard Bürdek's work (2015) is a reference. What matters is the logic of technological innovation and creativity as a social representation when configured with the user in mind—for example, new software, a toy, or some other object on the production line or recently released. The first is the new design of great social importance. It determines not only the differences between the models and the design itself but also, and above all, the differences between social classes. The subtleties of style and the redefinition of functions versus current technologies, which fit a concept of innovation—all of this and other elements involve the work of the designer. Design is also a game of power that necessarily encompasses the difference between social classes. At that moment, the creative designer's repertoire is confronted with its professional survival, which is the internal logic of capitalism—the profit motive. Apart from questions of political and ideological background, as part of the capitalist system, even unconditionally, the probability exists that such professionals, through their creative power, conceive of serial objects without the redundancy of their characters, their aesthetics, and even the quality to be differentiated. Therefore, the strength of the stratification of consumption and the internal logic of exchange relationships weighs on consumers. Similarly, there is the weight of purchasing power, which distinguishes classes in terms of the balance of consumption. Aesthetic innovation and the renewal process directly involve not only creation but also the significant turnover of consumption. Technological innovations and the desire for new objects encourage consumers to integrate further into the logic of mass society, in search of the visibility of others. Thus, the production and mass reproduction industry require a certain audacity from creators in their creative process. All social discourse is in a zone of deterrence, articulated by the ideologies of visibility, transparency, versatility, and consensus, to expose objects in a relationship that transforms cultural goods into consumer goods. Art for the object or the object for art is revealed in spaces without any aesthetic sense. Consequently, something is expected, which can be a new look or any other sense of innovation, with new representations of the old ways.

The importance of functionality that emerged with the Bauhaus later influenced globalized production, starting in the United States. However, this refers to the projects that are capable of dealing not only with the aspirations of the individual in society but also with the demand of a standard and mass industry. This situation took architecture and design to the ultimate consequences for the status of the obsolete object and pollution today, favoring the market for goods and consumption after the Second World War. Innovation has a short life in mass society, with a few notable exceptions. In a short time, a new product can become just another discardable gadget. However, considering theory and practice, the goal is the critical application and understanding of reflections on design in its technical essence. In this sense, it is a process in which it is possible to understand the influence of artistic and cultural creation in its representative aspects and empirical significance. Of course, the most considerable value is the direction in which designs have been taken concerning post–Second World War demands, globalization, and the relationship between the Fordist economic model and the current condition. Formal innovations no longer follow the function in its sole determining objective. Now, it is associated with consumers' identity values and meeting their needs and lifestyles, following criteria aimed at the sustainable development of diversified and fragmented production.

The criticism of the Bauhaus for having commercialized much of its production (e.g., furniture and objects) after the Second World War is unfounded but has continued to this day. Thus, that its products became objects of consumer society a little later is understandable. Now, no institution in capitalism will survive if it does not consider the principles of a system based on consumption and adapt accordingly. Consequently, the responsibility of design encompasses many considerations to succeed, both in terms of high commercial acceptance through meeting consumer desires and social needs. Today, it is about sustainable development and capitalism. Even philanthropic institutions that formally do not aim for profit must adapt to capital market rules. Conscious consumption is becoming necessary–meaning the recognition and responsibility to consume while preserving social and natural environments. In this sense, the leading philanthropic organizations are stimulating programs and partnerships. As a reference, the World Wide Fund for Nature (WWF) understands that working in partnership and collaboration with corporations enables effective solutions to the problems that threaten planet Earth due to the scarcity of natural resources, greenhouse gas emissions, accumulation of nondegradable materials, depletion of natural resources, and the model of a consumerist society of continuous expansion of production and sales, which is unsustainable in the long term. The leading solutions are recycling and the reuse of new renewable energy sources. Moreover, it is up to designers to project thinking about products' life cycle that prioritizes four factors: Nonpolluting and low-energy materials; functionality and maintenance optimization; reuse and recycling after disposal; and alternative design in the elaboration of ecological products according to environmental standards, involving products, packaging, marketing, and advertising strategies.

In the face of these production challenges and new consumption habits, these design capabilities alone are insufficient. The whole society must be interconnected. As soon as the first concerns about the scarcity of natural resources and other ecological consequences emerged in the 1960s and 1970s, Victor Josef Papanek, one of the most recognized designers to advocate for socially and ecologically responsible product design, emphasized in *Design for the Real World* ([1971] 2019) the agreement with the designer's responsibility to develop a product that is well received in the marketplace. However, Papanek made it very clear that responsibility must go beyond this consideration and be implemented long before design or redesign, questioning whether the product will be "on the side of the social good or not." He added the following: "Man has more basic needs than food, shelter, and clothing. We have taken clean air and pure water for granted for the first ten million years or so, but now this picture has changed drastically" ([1971] 2019, 66–67). Papanek's considerations resulted from his experience as a designer, mainly for a post–Second World War market, which led him to realize the destructive consequences for humanity. The relationships between designers, industry, and society were already being indicated in their responsibilities, and Papanek was outlining the first steps: "While the reasons for our poisoned air and polluted streams and lakes are fairly complex, it must be admitted that the industrial designer and industry in general are certainly co-responsible with others for the appalling state of affairs" ([1971] 2019, 67).

In this sense, the WWF has sought corporate partnerships and collaborations over the past decades. The WWF considers companies to be the drivers of the global economy; therefore, it deems them responsible for ensuring that the natural resources and ecosystems that make their businesses viable are used sustainably. The need for control to maintain this objective and achieve the goals established by environmental agreements, thus complying with the established principles, has come under the scrutiny of international institutions. Thus, the WWF emphasizes that for a company to ensure its legitimacy, it is necessary for it to practice sustainability by adopting environmental, social, and corporate governance; moreover, it must recognize this current market and issues and develop itself with goals and positive results in favor of the socio-environment. Even today, we can still register Behrens' legacy regarding the standardization of corporate identity. Therefore, it is worth highlighting that the critical thing is not only the aesthetic aspects of the architecture, environment, products, packaging, and logo but also the ethical aspects of the construction of the corporate image. Thus, it is no wonder that Behrens has been called the first industrial designer in recognition of his projects for industrialized products and the visual identity project. His work for the company Allgemeine Elektricitäts-Gesellschaft (AEG) is considered the first complete visual identity program. For Behrens, design was the basis of progress. One must consider the context of that time for the values to consider when building a "corporate image." However, considering the development, new values, and sociocultural transformations, this standardization not only implies the form following the function—that is, the aesthetic aspects respecting the functionality—but also implies the representation of the objectives of a company, its mission, and its values, which are mirrored in all of its representativeness and actions. As the market, societies, and cultures develop, transformations imply new measures and adaptations––design and redesign––in search of new conceptions, solutions, and undoubtedly innovations to keep the market system running. In addition, industrialists who produce in compliance with environmental laws have become certified by the International Standards Organization (ISO 14000). This seal is fundamental to corporate identity today. Sustainable development is development that is capable of meeting the current generation's needs without compromising the ability to meet the needs of future generations. Therefore, the green economy proposes, in addition to productive and social technologies, the creation of means by which essential factors linked to socio-environmental sustainability are not ignored in economic decisions with programs that promote environmental preservation or recovery, supporting segments of the population seeking an income from the recycling of solid waste.

Therefore, the green economy is to consume less and more consciously following initiatives aimed at rationalized products, which should be optimized for low energy costs or renewable energy and made from renewable and environmentally friendly materials. In addition, throughout all production processes, consumption planning, and product life, waste should be avoided. Thus, technological and scientific development in partnership with industries has intensified research into alternative energy solutions. In many countries, becoming less dependent on nonrenewable energy sources has become the new economic order. However, some countries still resist this demand based on political and economic interests. The most important treaties that condition decision-makers to follow sustainable development and a green economy are strongly supported by the United Nations (UN). As early as 1969, the International Council of Societies of Industrial Design (ICSID) made the design community aware of the need to pay attention to new design projects that promote quality of life concerning the quantity of production. The ICSID, later renamed the World Design Organization (WDO), is a nongovernmental organization founded in Paris in 1957 that protects and promotes the interests of the industrial design profession. Today, the WDO advocates design for a better world by promoting and sharing knowledge of industrial design-driven innovations that economically, socially, culturally, and environmentally enhance quality of life. In the face of globalization, imports and exports have been optimized by new communication and information technologies through the Internet, enabling connections and transactions of trade and production. In this sense, design is connected with art, technology, and society and, undoubtedly, attuned to the transformations and needs of its time as well as potential consumers, a contemporary population in its differences and similarities, which––since this population is a potential consumer target––is the basis for the conception of a new product or service. The conception of such a product should mainly consider what is most significant for contemporary urban life, such as democratic ideals and sustainable development. To this end, social responsibility is one of the primary characteristics of this creative activity in its functional, rational, and objective aspects. This is not to the detriment of the consumer's sensitive values, desires, and particularities of consumption, but rather in the synthesis of the relationship between reason and sensitivity to meet in, all its aspects, the aspirations idealized by society. This is precisely where the potential of design lies, in creating and recreating new habits and functions based on research and transdisciplinary associations to reach the optimal solutions for social inclusion. Moreover, cultural reflection is part of a natural transformation process of human beings and their creations. Today, new disciplines, such as design, have acquired greater prominence and an essential significance for developing a globalized society. Under these conditions, design and its respective formats as well as innovative ideas have reached a new status in empirical actuality accompanying technological evolution. Bürdek (2015), in this sense, made particularly explicit remarks when discussing the various definitions and descriptions of design, suggesting that instead of a new definition or explanation of design, we need to talk about specific problems that design must always solve. Among these problems, the author highlighted the context of technological change; prioritization of handling and easy use of products; optimization of production, consumption, and reuse; promotion and communication of services; and above all the avoidance, when necessary, of senseless products. The demand has been growing for professionals who work in the economic production sector, design each product, make its function visible, and enable a clear reading by the user. This specialty is known as product design, design and architecture, graphic design, and many other branches that have enjoyed recognition and even prestige. In addition, as is known, design must extend beyond the product itself and answer the questions of its essence, which is planning. In design, forms, functions, and meanings are not exhausted and, according to the most diverse contexts, cannot be independent because they would always be conditioned to the ongoing changes and perspectives of the object in the universe that surrounds it. Moreover, the design must respond to the environment and energy-saving and reuse issues. That is, the principles of sustainable development must be respected.

Equally, Bürdek (2019) made an important observation, precisely as the term "design" is being trivialized, by reminding us that the idea of Gropius––the founder and first director of the Bauhaus in Weimar––was to find a new synthesis of art and technology. However, this subject is known to be obsolete today. Then, in Germany, the meaning of design remains much closer to the initial term employed by Gropius: *Gestaltung*. Bürdek noted that even politicians state that they want *gestalten* (shape) in the sense of changing conditions, reality, social patterns, and societies. The notion of the social construction of technology refers precisely to this connection between technology and culture. In the sense of creation, the idea of social structure can play a central role in the relationship between technological development and cultural transformation. Economic issues accompany social issues and vice versa, and according to Bürdek (2019), Asian countries are bringing together high technology and design. For them, design is an essential tool for developing their economy and nation. Design is linked to technology, the economy, society, ecology, and aesthetics. Technology is the most critical, because it is where new concepts are developed. Designers are not the inventors of these technologies, but they do transform them into products. In other words, designers interpret new technologies into easy-to-use products that visualize the social values of their users. Accordingly, the relationships between technique, technology, design, education, and knowledge are inseparable. The definition of technique has its origin in ancient Greece (see Chapters 3 and 7). Today, the concept of technical generates the idea of a technological knowledge procedure––involving applied and industrial sciences. As a phenomenon, technology is also the subject of study in the applied social sciences and the arts. Since the Enlightenment, technology has been understood as the science of knowledge, which is the accepted definition today. Technology has been the main ally of knowledge and science in various fields of human activity throughout history and is the main objective of the analysis of changes in society.

The development of mechanization and industrialization, on the one hand, and the progress of science, on the other, have led to a complete transformation of the universe. In turn, technological innovation is continually being improved or modified by experts, simultaneously presenting both ephemeral and eternal values, which give rise to other innovations, always with new objectives to be achieved. The solutions, in this sense, are found through creativity, planning, and empirical research aimed at improved methods for the evolution of knowledge and development of society. Therefore, the meaning of the term "design" is linked to its close problem-solving objective. It is an essential part of the technological evolution to meet these new challenges for access to knowledge.

The Role of the Designer

The challenges in finding solutions lie in the process of the entire team involved in the project. However, it is not always possible to detect the causes of the problem so that solutions can be developed. Donald Norman, in *The Design of Everyday Things* (2013), reported some prevalent situations where knowledge-based problems are resisted under social and institutional pressures during product or service development: "Complex problem-solving is required when one is faced with knowledge-based problems" (Norman 2013, 345). However, it takes days for problems to be perceived. Furthermore, complex problems can generate new problems in the face of pressure for production and immediate results. Norman highlighted that "in commercial installations, the pressure to keep systems running is immense. Considerable money might be lost if an expensive system is shut down. Operators are often under pressure not to do this" (346). Moreover, understanding human behavior is highly complex compared with understanding systems with knowledge of physics or mathematics. On the one hand, solutions that attribute persistent problems to human error are no solution—on the contrary, this is to resist the possibilities of solutions. On the other hand, this complexity of problem-solving is found in all possible situations of human behavior. Therefore, it is necessary to apply some criteria in an attempt to get it right, thereby eliminating the possibility of error. Thus, Norman asserted that "root cause analysis is the name of the game: investigate the accident until the single, underlying cause is found" (306). However, one must find not the error but how the wrong decisions or actions were taken to prevent further errors. It is known that there can be many causes, not just one.

Furthermore, faced with Norman's statements, it seems that one can attribute human error not only in the use of a product or equipment (i.e., to the human user) but also in the design conception itself (i.e., to the designer's human error). Norman admitted the impossibility of solving problems if errors are not admitted. In his long experience as a designer, he reported that blaming people for an error makes it difficult to convince organizations of changes, reorganizations, and the need for redesign to solve problems. However, faced with this situation, Norman argued that "usually the system, the procedures, and social pressures have led to the problems, and the problems won't be fixed without addressing all of these factors" (313). In the face of this complexity, what role does design play today? Thinking about this complexity and the role played by the designer is to understand first from the whole and not from the parts and through a deconstruction of what is known to enable a reconstruction that seeks new ways to reach the correct result. The very construction of current knowledge has been guided by problem-based learning (PBL) pedagogy since 1974, which advocates identifying problems through observation, analysis, discovery, and recognition of the causes to eliminate them; furthermore, it involves verifying the effectiveness of the actions taken to eliminate the problem, comparing the situations executed (before and after). Next, a standardization is defined based on responsibilities, which is necessary to avoid new problems. Finally, all of the steps are reviewed for planning new projects, continually seeking improvements. Since 1999, industrial design has adopted design-based learning (DBL) in projects, which is based on the PBL model of three distinct stages for development. These stages are problem analysis, problem solution, and project report, focusing on the resolution of the problem through analysis.

As a result, DBL prepares the designer to visualize the design integration and, consequently, the innovation ideal, considering the alignment, research, mission, vision, concept, product, and process report. Thus, in seeking solutions to problems through these teaching methods, a crucial distinction stands out between problem-based learning and project-based learning. Whereas problem-based learning focuses on understanding, analyzing, and studying the problem, project-based learning focuses on the learning as problem orientation. Furthermore, project-based learning relies on developing projects based on the division and sharing of activities in a cooperative process. Thus, design, as a complex process in transformation, involves many human activities and skills related to the possibilities of projects through techniques and the constant stream of new resources made available by new technologies. The most significant challenges for designers lie in this scenario of novelties. Thus, the DBL methodology guides learning through creative ideas, skills, and competencies, aiming for new values in the face of information and knowledge (Stokholm 2014). However, one must consider that the design process—in all its ramifications—involves thinking as its main characteristic. The Cartesian consideration, based on the method of successive attempts to get something right based on mistakes, consists of finding flaws and consequently finding ways to correct them as solutions. However, the search for more adequate and efficient methods for evaluating and predicting problems in projects has become paramount when considering design as a social process project. Therefore, significant changes have been considered that involve the transferal of the problem-solving process to the problem-formulation process to meet a collective and successful purpose (Beckman and Barry 2007). Learning by doing is not new to these PBL and DBL methodologies, since the Bauhaus already employed the experience in its teaching methodology for its introductory course. Theoretically, this practice is found in the pedagogical theories of John Dewey. However, development and adaptation to the current context are highly relevant. Experience has been the focus in the training of designers. Specifically, the concept of experience is fundamental to detecting problems and solving them. According to Dewey, when we experience something, we act upon it and do something with it; then, we suffer or feel the consequences. We do something to the object of experience, and then it does something to us in return. Under such conditions, the action becomes an attempt; one experiences the world to know what it is like. Consequently, what one suffers becomes instruction––that is, the discovery of the relations between things (Dewey [1916] 1997).

Therefore, the designer's major challenge today is not only a methodological one but also a sociocultural one. The human intelligence challenge lies in the search for more knowledge and conditions us to review the central values of Western culture. Although humans have acquired autonomy in the modern world, guided by ethics or moral precepts for their existence, there has always been resistance to new values. Consequently, problems always arise in times of change and adaptation. This is something essentially human. In this sense, in the face of discoveries, theories of knowledge and pedagogy, and especially the development of science and technology, humanity is conditioned to the new paradigm, namely artificial intelligence (AI). The result of AI is an integral part of the digital transformation of all facets of our societies, including education and new designs and redesigns focused on sustainable development.

In 2019, UNESCO published a report titled *Artificial Intelligence in Education: Challenges and Opportunities for Sustainable Development*, which presented the following six challenges related to incorporating AI in education: (1) encompassing AI public policies for sustainable development; (2) ensuring inclusion and equity regarding AI in education; (3) preparing teachers to work with AI in education and preparing systems that meet educational demands; (4) developing inclusive and quality data systems; (5) developing meaningful research on AI in education; and (6) ensuring ethics and transparency in the collection, use, and dissemination of data. Under these conditions, in general, an awareness of the necessary processes is a determinant of the designer's work to meet the demands of sustainable development, with attempts to identify and solve problems. The basis of the sociocultural challenge is already found in the designer's training. It is also found in all social structures, especially in companies and industries–– where designers are essential––that cooperate with the programs developed by the WWF, as previously mentioned, which aim to achieve the UN's 17 Sustainable Development Goals (SDGs). Thus, design is related to sustainable consumption and production. According to the UN, this is SDG 12, which concerns "doing more and better with less. It is also about decoupling economic growth from environmental degradation, increasing resource efficiency and promoting sustainable lifestyles." The focus on "doing more and better with less" refers to an analogy of the design maxim "less is more," which has its origins in Behrens and was spread worldwide by Mies van der Rohe, as previously reported. The highlight in this sense is precisely the designer's responsibility in thinking about the consequences of their project, aiming at the elimination of any problem and, generalized in this case, meeting the objectives of sustainable development. Besides all of the other SDGs, "sustainable consumption and production can also contribute substantially to poverty alleviation and the transition toward low-carbon and green economies." SDG 9, which concerns inclusive and sustainable industrialization, is one of the fundamental challenges for design as it is centered on innovation and infrastructure, as discussed earlier in "Technicization and Aestheticization of Social and Urban Structures" (Chapter 8). This goal is fundamental as a stimulus for economic development, contributing to social inclusion being "a key role in introducing and promoting new technologies, facilitating international trade and enabling the efficient use of resources." In addition, regarding SDG 11 on sustainable cities and communities, as discussed in "The Urban Image in Motion and Connection" (Chapter 9), and reported by the UN, "the world is becoming increasingly urbanized. Since 2007, more than half the world's population has been living in cities, and that share is projected to rise to 60 per cent by 2030." This challenge also implies urban designers and architects' goal of planning cities for growth in a more sustainable and inclusive manner. The aforementioned three SDGs are highlighted to demonstrate where design acts directly; however, in sustainability's main characteristic of planning, one can also accept its indirect action with respect to the 17 SDGs, which are provided as follows for reference:

> (1) End poverty in all its forms everywhere; (2) zero hunger; (3) ensure healthy lives and promote well-being for all at all ages; (4) quality education, which enables upward socioeconomic mobility and is a key to escaping poverty; (5) achieve gender equality and empower all women and girls; (6) ensure access to water and sanitation for all; (7) ensure access to affordable, reliable, sustainable, and modern energy; (8) promote inclusive and sustainable economic growth, employment, and decent work for all; (9) build resilient infrastructure, promote sustainable industrialization, and foster innovation; (10) reduce inequality within and among countries; (11) make cities inclusive, safe, resilient, and sustainable; (12) ensure sustainable consumption and production patterns; (13) take urgent action to combat climate change and its impacts; (14) conserve and sustainably use the oceans, seas, and marine resources; (15) sustainably manage forests, combat desertification, halt and reverse land degradation, and halt biodiversity loss; (16) promote just, peaceful, and inclusive societies; and (17) revitalize the global partnership for sustainable development.

All social life activities are somehow related to the functions of design. Consequently, it can be understood that design must also assume a commitment to the 17 SDGs. Furthermore, much stronger cooperation of the entire global society is required to ensure the functioning of social structures, which depend on economic development as the main factor for making the SDGs possible. It is in this sense that ethical values are at the heart of the designer's work. The ethical notions that should be considered as core values of design are technology and innovation, along with their moral and political impacts on individuals and society. In this sense, the values should be considered during the design process as well as factors that influence projects, users, and third-party involvement in terms of functionality and its intended or unintended consequences. Ultimately, it is about awareness of the possible implications of design and the political and moral value choices that contribute to conscious design.

In the paper titled "Design for the Value of Sustainability" (2015), published in the *Handbook of Ethics, Values, and Technological Design: Sources, Theory, Values and Application Domains* (Hoven, Vermaas, and Poel 2015, 515), based on the definition of development aimed at the needs of the present without compromising future generations, Renee Wever and Joost Vogtländer questioned the viability of such sustainable development in the face of the global system. They stated that "additional requirements and objectives are needed to translate the meaning of sustainability to goals and requirements for designing products and services." The authors presented the steps of the design process in its complex relationship with the issue of sustainability, which is considered in each step as a criterion of purpose for completing the project, including determining the problems. They also stated that "relative outsiders, dealing with Design for Sustainability, should be aware of the different approaches and strive to align their organization's intentions with the appropriate approach and appropriate designers" (546). Consequently, the challenges for sustainability go beyond the basics of design in training, planning, and engineering. They comprise the reformulation of the entire system centered on economic development, environmental quality, and social justice. These changes are urgent and were discussed in the conception of "sustainable capitalism" in the work of John Elkington (1997), who highlighted that the challenges are related no longer to the triple performance of companies, industries, and economies, but rather to the complex challenge based on seven objectives. Elkington called these the "Seven Revolutions for Sustainability," which are as follows: (1) "market," where corporations will transform themselves and supply new markets in the face of the values demanded by society; (2) "values," aimed at sustainability as a political, economic, and social transition; (3) "transparency," through which even corporations that still operate outside of this complex demand will not escape this transition to sustainability; (4) "life-cycle technology," whereby activities and processes are supervised to maintain the value chain; (5) "partnerships," as a result of transfers of established markets and business relationships, namely the transformation of the competitive advantage of industrial sectors and companies; (6) "time," which concerns corporations that are unlikely to survive the sustainability transition over time; and (7) "corporate governance," which is a complex management involving many changes as the great challenge. These are just a general notion of the complexity surrounding the goals for a sustainable system, of which design is also a part. However, Elkington (1997, 353) still observed the following:

> Markets depend on rights and on responsibilities. To date, we have tended to focus more on our rights than on our responsibilities. This must change. In the sustainability transition, our own mind-sets will be crucial, whether we are acting, for example, as consumers, voters, parents, employers, or employees. And our mind-sets are strongly influenced by our education and training, by our experience, and by the information sources we use.

This is about the responsibility that involves the entire social structure highlighted in Elkington's analyses, which clarifies the relationship of this responsibility with the concept of "social capital."[5] In this sense, the term used in economics of "capital" is related to the sense of social structure in Bourdieu's (1986) conception. It is the sense of society in its possibilities of production and reproduction of sociocultural values that offers advantages for social mobility based on power relations and economic, social, cultural, and symbolic privileges––by diplomas and honorific titles. Thus, society's hierarchy also determines the power relations based on "social capital."

Accordingly, in the face of this complex relationship and objectives aimed at sustainability, when analyzing the basis of the designer's training and its relationship with the different contexts of industrial, economic, and sociocultural development in the passage from the 20th to the 21st century, it was first necessary to understand that the restructuring of the system has come a long way and still has further to go. Many transformations are required, which involve each small part of this complex field of productions and reproductions. The main points were highlighted regarding design challenges that concern planning and involvement with the entire social structure. Therefore, this construction of knowledge, identification of problems, and search for solutions as primary factors of the designer's social responsibility are focused on the ethical values of sustainable development. Consequently, achieving sustainable development through not only design but also all actions is dependent on the entire base of education and training of society. Mainly considering the inequalities in their reproduction by the social structures through education as "cultural capital," individuals must be enabled to have an education and knowledge that result in recognition and upward social mobility—meaning a position that is not only economically superior but that also allows leadership in decisions. Therefore, one of the main goals for sustainable development is the right to education as the basis of human rights, which consequently encompasses all other goals. Moreover, the interests are diverse but converge into two opposing camps, one of resistance and one of innovation. Consequently, as the innovations advance toward sustainability, this field of transformations gains more productive and reproductive strength, gaining more space in the system of "sustainable capital," mainly because this is the official rule and value for all humanity. As a result, everyone is directed to adhere to this orientation. There is no doubt that the precursors of this sustainable development are developed countries. However, this change depends on investments and support provided to emerging countries so that they can also have advantages, especially those exploited by their former colonizers and still are by forms of capital domination—and consequently by "sustainable capital." However, for humanity to genuinely succeed, agreements based on the 17 SDGs between major developed countries are essential.

To sum up, this chapter related sustainable development to the history of design in its relationship with planning, formation, configurations involving art, technique, and architecture in its industrial conception, as well as the socio-cultural consequences of its practice in the process of urbanization and technological development. An understanding of the term "technique" was sought in its origin, focusing on the passage from the 20th century to the present day, considering design in its significance in the search for solutions for society. The meaning of technique in its global dimension is fundamental for constructing equipment, machines, houses, clothing, and food, as well as for elaborating resources and techniques for teaching and learning by problem-solving and projects oriented toward identifying solutions to problems. The complexity of sustainable development lies in planning the entire production chain and urban lifestyle, not only in technological and scientific development but also, and mainly, in training and education. This is complex because of the condition of the social structures, including political-ideological aspects and beliefs. Besides all of the factors discussed, it is a context in movement and transformation–– sustainable development––that generates consequences that result in a back and forth, or evolutions and involutions, with a greater tendency to advance toward sustainability. Finally, through a theoretical foundation based on art, culture, and technological change, the importance of the SDGs and the responsibility of design concerning the global and local contexts in its potential for innovation and social transformation were recognized. However, it remains for the next and final chapter, titled "Aesthetic, Social, and Urban Solutions for a New World View," to discuss the essential aspects of climate change as part of everyday urban narratives and utopias that shape an uncertain future for human lives and societies. Utopias are essentially considered because, even if they are not realizable and remain as ideas, they correspond to a critique of reality. They are discussed regarding the imaginative conceptions contained therein, which somehow stimulate problem-solving and other meaningfulness realizations as solutions.

Notes

1 The origin of the word "design" comes from the Italian *disegno*, a concept that was used from the Renaissance onward for the purpose of designing, drawing, and representing ideas. In England, the concept of design began to be employed in the 16th century as planning for something with an object of art. Throughout design history, many definitions have been provided: What it would start with, then what function it would have, what domain it was in, and what its main points were. Today, however, it is no different. In the German language, until 1945, the word design was not used; there was instead *Produktgestaltung* (composition of the product) or *Industrielle Formgebung* (conception of industrial form).

2 Ludwig Mies van der Rohe stated the following in Detlef Mertins' monograph:

> I had to make a drawing for a facade for a factory. There was nothing to do on this thing. The columns were 5.75 meters (19 feet). I will remember that until I die. I showed him a bunch of drawings of what could be done and then he said, 'Less is more', but he meant it in another way than I use it.
>
> *Mies* (Phaidon Press, 2014)

3 Isolde Karen Herbert:

> Yet Morris does not ask for a retreat into either nature or the past. In fact, as May Morris remembers, her father considered the back-to-nature philosophy of the

> "simple-lifers" (CW 22: xxiv) untenable because a retreat from modern complexities is both escapist and impossible. As for the past, Morris asserts that he is not "a mere praiser of past times" (CW 22: 163) or "a mere railer against "progress" (CW 23: 281). True progress is not a reconstruction of the past (after all, this is what Anti-Scrape opposes) but an awareness of the contribution of the past to civilization as it could/should be.
>
> *Journal of William Morris Society*, v. 10, no. 1 (Autumn 1992): 4–9

4 Isolde Karen Herbert:

> Commerce mocks humanity's traditional reverence for the earth: instead of inspiring the desire for beauty, commerce inspires the desire for profit. Consequently, in "The Revival of Handicraft" (1888), Morris suggests that our attempt to produce beauty for its own sake is now a futile imitation of the originally innate instinct for beauty––futile because, as our abuse of the environment indicates, we lack the "eyes" to see nature comprehensively; instead, we perceive nature reductively as "raw material" or "property."
>
> "Nature and Art: Morris's Conception of Progress."
> *Journal of William Morris Society*, v. 10, no. 1 (Autumn 1992): 4–9

5 According to the bibliographical references, this concept is understood to have been introduced by Pierre Boudieu (1979, 1980) in comparison with James S. Coleman (1988).

Bibliography

Beckman, Sara L., and Michael Barry. "Innovation as a Learning Process: Embedding Design Thinking." *California Management Review* 50, no. 1 (October 2007): 25–56. https://doi.org/10.2307/41166415.

Bourdieu, Pierre. "Les trois états du capital culturel" [The Three States of Cultural Capital]. *Actes de la Recherche en Sciences Sociales* 31 (1979): 3–6. https://doi.org/10.3406/arss.1979.2654

Bourdieu, Pierre. "Le capital social, notes provisoires" [Social Capital, Provisional Notes]. *Actes de la recherche en sciences sociales* 31 (1980): 2–3.

Bourdieu, Pierre. "The Forms of Capital." In *Handbook of Theory and Research for the Sociology of Education*, edited by J. Richardson, 241–258. New York: Greenwood, 1986.

Bürdek, Bernhard. *Design. The History, Theory and Practice of Product Design*. Basel: Birkhäuser, 2015.

Bürdek, Bernhard. "To What Extent Is Bauhaus Even Possible Nowadays?" Interview by Christiane Wagner. Special Edition, Art *Style, Art & Culture International Magazine* v. 2, no. 2 (June 2019): 85–91. https://doi.org/10.5281/zenodo.4088827

Coleman, James S. "Social Capital in the Creation of Human Capital." *American Journal of Sociology* 94 (1988): 95–120. http://www.jstor.org/stable/2780243.

Dewey, John. *Democracy and Education: An Introduction to the Philosophy of Education*. New York: Macmillan, 1916. Reprint, Project Gutenberg, 1997. https://www.gutenberg.org/ebooks/852

Elkington, John. *Cannibals with Forks: The Triple Bottom Line of 21st Century Business*. Oxford: Capstone Publishing Limited, Oxford Centre for Innovation, 1997.

Herbert, Isolde Karen. "Nature and Art: Morris's Conception of Progress." *Journal of William Morris Society* v. 10, no. 1 (Autumn 1992): 4–9.

Hoven, Jeroen V.D., Pieter E. Vermaas, and Ibo van de Poel, eds. *Handbook of Ethics, Values, and Technological Design: Sources, Theory, Values and Application Domains*. Heidelberg, New York and London: Springer, 2015.

Meggs, B. Philip. *A History of Graphic Design*. Michigan: John Wiley & Sons, 1998.
Mertins, Detlef. *Mies*. New York: Phaidon Press, 2014.
Norman, Donald A. *The Design of Everyday Things*. Revised and Expanded Edition. New York: Based Books, 2013.
Papanek, Victor. *Design for the Real World: Human Ecology and Social Change*. 1971. Reprint, New York: Thames & Hudson, 2019.
Stokholm, Marianne. "Problem Based Learning versus Design Thinking in Team Based Project Work." *Proceedings International Conference on Engineering and Product Design Education*, University of Twente, September 4–5, 2014.
United Nations. "Sustainable Development Goals." Accessed June 8, 2022. https://www.un.org/sustainabledevelopment/sustainable-development-goals/
Vogtländer, Joost, and Renee Wever. "Design for the Value of Sustainability." In *Handbook of Ethics, Values, and Technological Design. Sources, Theory, Values and Application Domains*, edited by Van den Hoven, J., Pieter E. Vermaas, and Ibo van de Poel, 513–549. New York: Springer, 2015.
Werkbund Archive. "Museum of Things." Accessed June 10, 2022. https://www.museumderdinge.org/institution/historical-key-topic
World Design Organization. https://wdo.org
WWF (World Wide Fund for Nature). https://www.worldwildlife.org

12

AESTHETIC, SOCIAL, AND URBAN SOLUTIONS FOR A NEW WORLD VIEW

Discussing the relationship between art and nature is first necessary to contextualize them within the origin of the history of humanity, cultures, and civilizations, and therefore, within the development of science, technology, and progress. However, social, political, and economic aspects also determine this discussion. These aspects are based on perceptions of art and nature in their interdependent relationship and association with progress. This discussion seeks to understand the meaning of an environmental aesthetic, considering that many approaches exclude the artistic object, which is fundamental to philosophical and aesthetic reflection. Art has its values, both aesthetic (*aisthésis*) and technical, and appreciates other values only in the function itself. The disproportion of aesthetic (*aisthésis*) and nonaesthetic (*anaisthésis*) values is often associated with sociocultural contexts. Nevertheless, contemporary aesthetics in its context encounters a problem in its interaction with science and ethics due to the necessary specific logic and new criteria in art. Furthermore, this discussion seeks answers to the difference between artistic and natural appreciation and an understanding of media's aesthetic effects and resonances across socioterritorial boundaries. In theory, an agreement for combatting climate change in terms of the world's industrial and financial aspects through communication technologies and artistic practices could integrate art with a social critique through the influence that images exert, as presented in Chapter 11 "Designing for Sustainability." However, such an agreement could present images of a nostalgic return to nature artificially, valuing biodiversity transmitted by visual arts and thus based on out-of-context facts. In this sense, two aspects are essential for aesthetic reflection concerning context and cultural transformation: The first is technique, under which the term "art" is included, and the second is knowledge based on perception for verifying facts when considering the effects of images. These images appear in contemporary cultural events and media and allude to everyday life in natural and urban spaces,

DOI: 10.4324/9781003314943-13

toward environmental and everyday aesthetics. In addition, they raise discussions and present new elements of the relationship of the media image. Reflections are sought to appreciate natural, human, and, in general, urban environments with natural and urban aesthetic appreciation (Berleant 1984).

Accordingly, this approach is outlined in three parts. The first, titled "Weltanschauung, Utopias, and Environmental Aesthetics," refers to the historical background of the early impacts of the technological revolution on urban life, and consequently, on the values of a "romantic nature" in opposition to industrial development. In this respect, the bases for a worldview and utopias concerning a "back to nature" sense are presented with origins in European Romanticism. In the second part, titled "Nature, Culture, and Reality," the environmental scenario of sustainability is analyzed in its economic dimension, focusing on the social context of the emerging political agenda as the main challenge for corporations in the 21st century, which respectively concerns the challenges of architecture, design, and urban planning. The primary basis of this analysis is the meaning of "sustainable capitalism." In the last part, titled "Cities' Pollution: The Ethical and Aesthetic Impacts," the Sustainable Development Goals (SDGs) are discussed in terms of their complexity. The responsibilities of planners in design and architecture, especially for urban design, lie in planning for long-term purposes. This planning depends on the ethical and aesthetic aspects of public policies and the population's needs focused on solutions to urbanization. Finally, this chapter's discussion concludes with the new worldview based on the goals and orientation of the United Nations (UN) as the current context—the *Zeitgeist* and leitmotif for awareness of working together to meet the SDGs.

Weltanschauung, Utopias, and Environmental Aesthetics

Due to scientific and technological advances in the face of climate change, the world and people's view of it—Weltanschauung—are changing. However, according to Timothy Morton (2007) in *Ecology without Nature: Rethinking Environmental Aesthetics*, ecological consciousness requires a shift in vision from anthropocentrism to ecocentrism. Morton emphasized that the idea that a vision can somehow transform the world is profoundly grounded in the Romantic period and the notion of world vision—Weltanschauung—itself. Consequently, possessing a new vision of the world implies dealing with how human beings experience their place in the world. In this manner, aesthetics plays a determining role, establishing patterns of feeling and perceiving that place in the world. Therefore, by analyzing the configurations of society throughout the evolutionary history of humanity, it becomes apparent that technology transfer has always been an interaction between nature and culture for all possible achievements of human civilization, as Wolfgang Welsch (2004) stated concerning Darwin's evolutionary aesthetics. Thus, climate change and other complex environmental crises can be understood under aesthetics analysis with consideration of digital media and the processes of communication and image archiving as memory.

Moreover, in contemporary society, reality is thought to be related to the appearances or representations that shape the abstract and concrete manifestations of the truth (i.e., the universe of appearances is believed to represent a logical process), much like Hegel's worldview in *The Phenomenology of Spirit* (1807). Therefore, in Kant's theory conception of the Transcendental Aesthetic, the dynamic established between appearances and objective reality pertains to the relation between the object and space-time. When a thing can be perceived by one or all of the senses, it is understood and finally conceptualized and defined as reality. An image can be interpreted in many ways at the level of communication or artistic creation because of the ambiguousness of the sense in which it is transmitted. In today's discussion on environmental issues concerning "the green image," the phenomenological approach is fundamental. Along this line, Morton stated that "people may eventually recognize in the Romantic period the beginning of 'environmental' ways of understanding and acting. Romantic history emerged, evoking the spirit of the age, the *Zeitgeist*, the flavor of a culture at a particular moment in time" (1997, 286, my italics). Morton further added some relevant observations relating the context of Romanticism to our recent past, observing that ecology is also a form of awareness of belonging in the world by exemplifying that "in material historical terms, environmental phenomena participate in dialectical interplay insofar as they bring an awareness of environmental negatives such as global warming" (289). In addition, contributing to this discussion are the author's remarks about the fact that Romanticism's art and philosophical theories continue to be maintained in contemporary ecological narratives. In this sense, the sociocultural, political, and economic contexts are in need of updating.

Referring to Romanticism does not mean going backward or involuting. On the contrary, when dealing with a period that marked the literature and art of the late 18th century and part of the 19th century, the most significant matter is precisely the worldview and the strength of an original thought manifested in the entire cultural, social, and political environment. This is contrary to the rational and technological rise engendered by capitalism. Among the most significant references to the romantic utopia is the work of Walter Benjamin. In his 1927 essay, originally titled *Traumkitsch* (oneiric kitsch), kitsch is the side through which the thing offers itself to the dream. Benjamin presented a sense of the dream's participation in the sociohistorical context, conditioning limits or even reason on the invention. In this regard, Benjamin was aware of the novel by Georg Philipp Friedrich von Hardenberg, known by the pseudonym Novalis, titled *Heinrich von Ofterdingen*, which was of great importance in Germany during Romanticism. The first part of the novel narrates a dream that the poet of the High Middle Ages, Heinrich von Ofterdingen, had about the "Blue Flower." In the dream, the poet follows a path that takes him to an unknown world, passing through the practical experiences of life as in a tale. On the way, he returns to the origins of knowledge and, in the end, finds the Blue Flower. An eremite informs him that the art of writing history is not about simple records but rather, and always, about an interpretative project of configuration (*Gestaltung*).

The Blue Flower is of great symbolic relevance as it refers to longing, love, and the metaphysical ambitions of the infinite. Thus, Benjamin began and built his reasoning in *Traumkitsch* (oneiric kitsch) regarding the relationship of objects with humans, aiming at a search for meaning and their original value in a different condition. Under such a condition, dreams would no longer follow the infinite paths of time and space to meet the knowledge of nature, human, and spirit, but would rather follow the ways of everyday life—of banality—in modern times. In this analysis, Benjamin's reflections were attributed to the phenomenological aspect of Hegelian aesthetics, the presence of a sociohistorical philosophy concerning poetic images, which evoke mental images as the Romantics did, who loved freedom and perceived the power of nature over humans—the sublime. Furthermore, the Romantic *Zeitgeist* influenced the entire sociocultural sphere. Nature was always related in its dimension of power over human beings and mysteries. Today, many references are attributed to this era in environmentalist narratives considered utopias. However, as previously discussed in "Technicization and Aestheticization of Social and Urban Structures" (Chapter 8), utopias are essential as a critique of reality. A utopia presents the imagination of an idea based on an analysis of social conditions.

Consequently, even if these ideas are not realized, they can awaken the conscience for many other similar realizations and improvements to society—to humans' quality of life—approaching ideal solutions that, with time, can be realized according to the experience and knowledge acquired. For example, in addition to his artistic and literary achievement, Goethe thoroughly investigated the original plant hypothesis (*Urpflanze*), which in a way refers to the idea of the Blue Flower. However, despite the evident appeal of this morphology for ecologists, Goethe's use of this concept followed the Neoplatonic and idealistic tradition typical of an antiscientific, utopian, and illusory romanticism. Still, it was a way in which he stimulated research and discoveries by science. In this sense, in the face of global warming, scientists have for long studied the resistance of plants in the face of drought to more accurately evaluate the consequences of climate change for the biodiversity of specific ecosystems. In Romanticism, the painter most influenced and admired by Goethe, among the Germans, was Philipp Otto Runge (1777–1810), who associated the metaphysics and symbolism of colors in his works. The best example is perhaps even the leitmotiv of the Blue Flower, which remains alive and continually represented in German culture, especially in literature and music, from Goethe to Franz Schubert (Schmitz 2008). Therefore, essential to the discussion of significant theories on the reflection of the relationship of Romanticism with representativeness in visual narratives, kitsch—analogous to the symbol of the Blue Flower—concerns the leitmotiv of the Romantics from Novalis' novel. It does so through the representation of ambiguous meaning, of the final experience with the infinite and historical interpretation, with allusion to fantasy, dream, and Benjamin's oneiric kitsch—an experience not lived and compensated by appearances concerning nature versus culture.

However, in the contemporary purposes of a "green image," especially concerning urban centers, the developed narratives somehow present ambiguity and utopia. On the one hand, written and visual narratives present the destruction of nature, the idea of the end of Planet Earth, and the idea of the "Anthropocene" as described in "The Arts' Function: Social Inclusion and Sustainability" (Chapter 10). On the other hand, many images, landscapes, and architectural environments focus on appreciating natural spaces as a reflection of urban development, in the romantic sense, against the characteristic of large cities of artificiality, seeking a return to nature. However, this artificiality remains as the spaces and images become landscapes configured in the sense of the romantic picturesque, that is, of an aestheticized nature. These observations concerning the romantic ideal and narratives focused on environmental and ecological movements are currently still in the aesthetic model of Romanticism. The foundations lie in the relationship between beauty and the sublime and in the theories of Immanuel Kant (1790) and Edmund Burke (1757). Burke presented the sublime in aesthetic experience, emphasizing the essential foundation of this feeling for the imagination. Burke's ideas influenced Kant, especially his observations on the feelings of beauty and the sublime.

To Kant, the feeling of beauty is about pure beauty, which refers to the aesthetic judgment without finality, understood as a free contemplation of beauty. According to him, the natural environment offers, *a priori*, pure aesthetic judgments with no intention of meaning. It does not refer to any object or concept. On the other hand, however, applied beauty presents the ideas of finality, applied taste, and being better concerning the ideal model—a vision of perfection. For example, a natural landscape painting could feature flowers with brighter colors and more vibrant greenery than the artist originally perceived. This meaning lies precisely in the picturesque idea, which emphasizes the aesthetic experience of nature and its connections over time between the aesthetic appreciation of nature and its treatment in art. The picturesque theory advocates aesthetic appreciation in which the natural world is experienced as if it were divided into art scenes, ideally resembling works of art, especially landscape paintings. This idea also resonates with other artistic traditions of nature as a reference for art in a mimetic process. This concept originates in nature as the object of aesthetic experience offering a natural and superior beauty to that of art, and also in the development of the concept of disinterest, that is, a sole focus on the aesthetic appreciation of nature without the personal, religious, economic, or practical interests of the appreciator. Nonetheless, disinterest in aesthetic experience guided by nature's beautiful and sublime forms resulted in the picturesque. Picturesque "literally means 'picture-like' and the theory of the picturesque advocates aesthetic appreciation in which the natural world is experienced as if divided into art-like scenes" (Carson 2020), which are like landscape paintings.

The finality of the communicated opinion of taste matters in the aesthetic judgment itself. It concerns the purpose and not the content. However, the difference between natural and artistic beauty—that is, between nature and art—can be compared with the feelings of the beautiful and the sublime. The relationship between nature and art is mainly due to the difference between them as well as human beings' constant search to overcome natural beauty—pure, without purpose, without interests—using artistic beauty, which through techniques and artifices aims for a purpose—amid the diversity of interests—and a perfection. Overcoming natural beauty was precisely the purpose of the fine arts. According to Kant, realizing a work of art that characterizes nature as its origin, aiming for the highest formal quality in presenting it as a work of art, is not a product of nature but rather freedom from any natural imposition. It is the result of subordinating imagination to understanding; valuing the harmony of the faculties but without establishing any interest, objective, or pretension; and enabling an acceptable universal communication, above all without relying on concepts. However, the notions do not concern an imitation of nature, simply because if imitation were a rule of beauty, this rule would be established as both concept and purpose. Then, all works resulting from the imitation of nature would be beautiful. Each one would have its definition of beauty, of beauty for beauty's sake, without establishing sufficient reasons for the satisfaction of the ideal of beauty, thus conditioning the representation—the imitation—to an unsatisfactory finality. There is no concept of beauty. The feeling of beauty searches for knowledge and truth, and therefore, it aims for pleasure, beauty, and the good. The beautiful and the sublime are individual feelings and imply some purpose when expressed, seeking universality—agreeing with common sense. Beauty's pleasure is defined by the harmony of our capacities for understanding and imagination. The reference is nature. Work close to nature, a natural work, offers nothing exceptional to the senses because nature is objective and neutral and serves as a reference or model. According to Kant, the intense feeling of infinity, of a force that is not under our control, is the intuition of the idea of infinity, of the immeasurable, which the imagination cannot reach, leading to the feeling of the sublime. It is the awareness of powerlessness before a natural force. In this sense, the superior power of nature over humans has been appreciated aesthetically in the work of art. During Romanticism, many paintings represented the sublime.

On the one hand, the picturesque as an aesthetic experience has its basis in the aesthetic theories of the 18th century from Burke and Kant. Throughout the 19th century, however, it lost its meaning. On the other hand, in practice, one can still notice aspects of the picturesque as an aesthetic experience when art is idealized with nature as a reference, especially in mediating images that allude to nature aimed at tourism, housing, and even parks and artificially natural environments, including open-air museums. One must also assume that the practical dimension of this conception extends to various contexts and places. Nature as a reference for the picturesque has expanded into an aesthetic of everyday life, and the environment has become a picturesque work. Today, even a work of art resulting from natural beauty can be a medium for a disinterested aesthetic and result in the picturesque. Examples are the painting *Le Jardin de l'artiste à Giverny* by Claude Monet (1900) and a painting by Thomas Cole, characterized by the realistic and detailed portrayal of natural landscapes in their beautiful and sublime aspects for the design and realization of residential gardens. To some extent, all appreciation of nature and artistic representation focused on a disinterested aesthetic would be conditioned to nature without human intervention. However, human participation becomes the main interest in the picturesque sense of natural and urban environments encountering an everyday aesthetic, with an emphasis on the work *Aesthetic Participation and the Urban Environment* by Arnold Berleant (1984), with an aesthetic committed to the aesthetic appreciation of not only nature and art but also all other aspects of the human experience. However, human awareness of nature did not emerge in Romanticism. On the contrary, human beings have possessed a sense of nature since their origins and throughout the entire history of society's development. However, the refinement of human preferences through techniques and knowledge is part of constant development, distancing humans from nature by the artificial and shaping distinct ways of feeling and experiencing the reality of things.

Nature, Culture, and Reality

The development of our cities is nothing more than the organization of how humans in their collective habitat seek improvement, progress, evolution, and continuity by creating, presenting, producing, and reproducing worldviews and knowledge formation. The origin of this organization was found 2.5 million years ago in the first phase of the Paleolithic period, when hominids moved around according to the conditions of natural phenomena, which characterized the nomadic life in groups at this stage of human evolution. They survived by hunting, fishing, gathering fruit, and—most vitally—discovering fire. However, a second phase represents a much more critical breakthrough for the development of the first communities, namely the end of nomadic life and the development of civilization and cultures until the current stage. This evolution depended not only on the discovery of fire but also exclusively on its control. Fire made life a much greater possibility. In addition to heat, fire protected hominids from being attacked by animals and enabled them to roast meat; moreover, social interaction began to solidify during this Middle Paleolithic stage. Imagine them gathering and relating to each other around a fire, as can be seen, for example, in Jean-Jacques Annaud's film *Quest for Fire* (*La Guerre du Feu*, 1981), which was based on a book written by J. H. Rosny (1911).

Moreover, by the upper Paleolithic period, hominids had gained extensive experience and created techniques for survival and mastery of the environment in which they lived. Henri Bégouën (1939) discovered and documented some of the earliest techniques of the upper Paleolithic period, contributing to new studies and an understanding of symbolic and rational thinking in the interaction of art and technique. Moreover, Birgit Recki (2021) recently approached technology and the arts by overcoming the ontological duality between nature and culture. This historical recapitulation is essential for situating our current condition and understanding that fire represents a vital role in human life. Consequently, alternative energies are new means for the continuity of humans, considering all of the fragilities that humanity encounters in the face of nature. Human evolution is primarily a consequence of skills and discoveries aimed at protection, survival, and domination of the environment.

Although the notion of the first communities is paramount for understanding the extent and condition of the development and evolution of cultures and societies, the effects of human actions on climate change are found at the beginning of the Industrial Revolution. Consequently, all phases of this industrial development, especially with the acceleration after the Second World War, represent a threat to the conditions necessary for human life on Planet Earth. This threat is caused, according to scientists, by humans themselves through exceeding the limits of their actions on nature: "Four of these boundaries have already been crossed—climate change, vegetation cover, biodiversity loss and extinctions (the Sixth Extinction) and biogeochemical flows—with phosphorus and nitrogen cycles playing a particularly crucial role."[1] The meaning of this development and its consequences on the environment have already been clarified in previous chapters. The current state of the Fourth Industrial Revolution, and the primary motivating force, which is the financial one directly related to scientific and technological development, have been following Cartesian reason.

Thus, from the mastery of fire, the end of nomadic life, the first communities (life in groups for safety and protection), the development of agriculture, the social distribution of labor, and rupestrian art to the first ritual manifestations to the present day—this is a quick review of the evolution of humans, a great leap for humankind. It is reminiscent of the scene in Stanley Kubrick's *2001: A Space Odyssey* (1968), which starts from a piece of a skeleton and ends with a spaceship, depicting the evolution from humans using bones as tools to modern beings. However, this simple but necessary retrospection also demonstrates the challenges facing our existence under the concept of the Anthropocene.[2] This concept appeared in the 1950s and produced controversies. The denomination is questioned by historians and anthropologists, who specifically question the reference of "anthropos" as a generic form of the human being and ask the following: "Who is responsible for crossing the biogeophysical boundaries if not Western humans and a particular socioeconomic system?" While these discussions are ongoing, some alternatives are emerging, such as the "Occidentalocene" and the "Capitalocene."[3] Nevertheless, to discuss the current context of the Anthropocene and the utopias of sustainable development, John Elkington's (1997) work *Cannibals with Forks: The Triple Bottom Line of 21st Century Business* addresses the challenges of sustainability, not just changes in climate but also system-wide changes to "sustainable capital." Most intriguing, besides the title, are the author's first descriptions in the book's introduction:

> "Is it progress," the Polish poet Stanislaw Lee asked, "if a cannibal uses a fork?" I believe it can be, particularly in the case of corporate capitalism and

> corporate cannibalism. If this last phrase seems far-fetched, read this description of Microsoft's founder, William Gates III, 'Bill Gates eats competitors with the methodical determination of a corporate Pacman's Gates, it is true, is scarcely renowned for his environmental or social sensitivities. But in our rapidly evolving capitalist economies, where it is in the natural order of things for corporations to devour competing corporations, for industries to carve up and digest other industries, one emerging form of "cannibalism with a fork"—sustainable capitalism—would certainly constitute real progress.

Indeed, there is no doubt about how the capitalist system works today and that its survival depends on commodification. In this system, everything, or almost everything, becomes a commodity of the liberal market, including culture, as discussed in the initial chapters. However, under these conditions, in Elkington's conception, sustainability plays the role of a tool compared with a fork. To understand the meaning of *Cannibals with Forks*, the allusion to "the fork, sustainability's triple bottom line," Elkington (1997, VII–X) clarified that "its three prongs are economic prosperity, environmental quality, and social justice." Elkington's work deals with the "seven revolutions" transforming the world, specifically the business world. However, this implies the whole system of life. Nonetheless, he stated that sustainability's triple bottom line "help[s] drive major corporations and leading economies towards these goals [sustainability]." The author accepted that, "while corporate Cannibalism may not be pleasant to behold or comfortable to experience, it will remain an intrinsic part of any competitive economy." He also believed that the challenge for these corporations lies in how to adapt and *sustain* the set of values to enforce sustainability. He posed the following question: "Is it possible, for example, to evolve new types of corporation which are less inclined to operate as economic, social, and ecological predators?"

Although Elkington directed his arguments to the environmental dimension of sustainability without delving into the social or economic dimension, he clarified the social importance of the emerging political agenda and the main challenges for corporations in the 21st century. Accordingly, based on the triple bottom line for sustainable capitalism as a tool—the fork—it becomes clear that good political will is required to focus on social sustainability's implementation. Nevertheless, in allusion to the cannibal with a fork (the devourer), who would be devoured? Would it be the less adapted in the face of natural selection, as in the ecological system? For example, with the variations in the natural habitat of polar bears as a result of global warming and lack of food, cases of cannibalism have increased in this species. Thus, in Darwin's conception, natural selection involves the environment according to the characteristics of the living beings most adapted to survival. Those who are more adaptive may develop, evolve, and transfer their resistance to their successors, strengthening the continuity of the species. Under these conditions, moving this theory to social life, the following question arises: How can we solve the problems of inequality through social justice so that less apt people are strengthened?

The political agenda for sustainability was developed geopolitically by the more able. However, the long-term goals are hoped to be achieved through the agreements presented by the UN Conference on Environment and Development (UNCED).[4] At the 1992 UN Conference, heads of state and government from all over the world approved a set of principles on the environment and development. Their aim was to lay new foundations for the production and distribution of the wealth generated by human work, which would contemplate the appropriate use of the resources offered by the planet and ensure everyone the right to live with dignity, both now and in the future. The document, known as Agenda 21, highlighted paths and defined the responsibilities of each social agent in the search for sustainable development. Agenda 21 was held in Rio de Janeiro and became known as Rio92 or ECO92, and the maxim that emerged from this meeting was "think globally, act locally."

Ten years later, at a new World Summit on Sustainable Development in Johannesburg, heads of state assessed the progress made. However, ecological threats had taken over the world scene since the 1980s. Hence, several conferences were held in Canada and Sweden on climate change, as already presented in "The Arts' Function: Social Inclusion and Sustainability" (Chapter 10), in the international representation of a commitment to the global responsibility of each country to reduce pollutants, specifically greenhouse gases. However, recall that these discussions were only negotiated in 1997 in Kyoto, Japan. Recently, the Paris Agreement, negotiated during the COP21 (UN Climate Change Conference 2022) in Paris, was approved on December 12, 2015 and officially entered into effect on November 4, 2016. In Glasgow, at COP26 (UN Climate Change Conference of the Parties) in October and November 2022, the targets were finalized by the delegations, which meant implementing the rules established in the Paris Agreement to ensure compliance with limiting global warming to 1.5°C. The Intergovernmental Panel—the UN body for assessing the science related to climate change—released the leading information about global warming in *On Climate Change*. In addition to all of the relevant causes, the report mentioned that "an increasing share of emissions can be attributed to towns and cities." Accordingly, "cities consume 78 per cent of the world's energy and produce more than 60 per cent of greenhouse gas emissions . . . they account for less than 2 per cent of the Earth's surface" (UN, Climate Action, 2022).

Cities' Pollution: Ethical and Aesthetic Impacts

Nothing is more important than energy as a source of life. Yet, how many forms of energy are human life and its creations as extensions of their existence dependent on? In the beginning, the only energy source was food—the very biological condition of humans and animals for muscular strength. This force was the model for many devices and energy systems that evolved, gave rise to new energy systems and sources, and above all contributed to the growth and progress of cities. From wood-generated fire to fossil-fuel fire, the pleasure of electricity, and all of the creations that provided comfort to human beings to the values of life in society, humans now face the problems of nonrenewable and renewable energy resources. Yet, all seem to converge toward common sense—the renewal of energy maintaining the pace of cities and economic system running.

Throughout history, many techniques and technologies have been developed to generate energy, which has evolved systems and are still used today, such as the power of wind and waterfalls, for hydraulic and wind energy sources. These energy sources are renewable, and there are also other forms such as biofuels, solar energy, ocean energy, and geothermal energy that depend economically on many technological resources to be viable. On the one hand, nonrenewable sources will not always be available to be extracted from the environment; moreover, they have an enormous impact on the environment through emissions of carbon dioxide and methane (i.e., greenhouse gases), leading to global warming, as well as severe consequences from problems involving forms of production and raw materials. On the other hand, renewable sources also have impacts on the environment; for example, the construction of hydroelectric dams in the Amazon presupposes the construction of many barriers, which consequently affect the habitats of indigenous people, river resources, and the displacement of urban and rural people at a high human cost.[5] Hence, the impact of these renewable sources causes an ecological imbalance in the region. There are also problems with nonsustainable agriculture and farming. As a consequence of that practice, in the face of the oil crisis of the 1970s, for example, Brazil aimed to develop renewable energy by investing in the production of ethanol, a biofuel extracted from sugarcane and widely adopted in the national territory. However, in 2008, Jean Ziegler, former UN Rapporteur on the Right to Food, reported that the mass production of biofuels would compete for fertile land and reduce the areas dedicated to agriculture for food, contributing to a rise in food prices. Therefore, it should be considered that sustainable development includes the objective of fighting hunger. More accurate information can be found in the annual reports during the period when Ziegler (2008) was a special rapporteur on the right to food to the Commission on Human Rights. The following information from the report stands out[6]:

> Increasing the production of biofuels could bring positive benefits for climate change and for farmers in developing countries, including improving food security, if the benefits trickle down. However, it is unacceptable that increasing production of biofuels should lead to greater hunger. The greatest risk is that dependence on the agro-industrial model of production will fail to benefit poor peasant farmers and will generate violations of the right to food.
>
> *(2008, 21)*

However, this observation, according to the information released by the press by Raymond Colitt (2008), who covered the event, Ziegler was referring to the United States' production of corn destined for biofuel production and not specifically to the Brazilian production.[7] However, when checking the report, in general terms, this problem should be given full attention so that the solution to one problem is not the reason for another. In addition, the Latin American reality is highly different from the reality in the United States. In Brazil, the production of fuel alternatives leads to an expansion of the labor market, which is a possible solution to the problems of emerging countries, as well as in all Latin America, besides reducing the negative impacts of the use of nonrenewable energies. In this sense, the words of Luiz Inácio Lula da Silva, Brazil's president in 2008, are equally timely and enlightening: "The real crime against humanity is to discredit biofuels a priori and condemn food-starved and energy-starved countries to dependence and insecurity" (Colitt 2008). When developed and emerging countries look for alternatives to renewable energy, this is a way to decrease the dependency on nonrenewable energies—an energy source produced from fossil fuels that predominate worldwide—and to provide economic independence. With the production of these alternative energy sources, the energy dependency of the current world model is reduced. Specifically, in the case of Brazil, not only does ethanol derived from sugarcane stand out but also biodiesel derived from vegetable oils. As the second-largest ethanol producer, Brazil is thus one of the countries that stands out for the use and production of biofuels. However, the country must manage plantation areas to avoid problems related to deforestation. In this sense, the report[8] presented the following note:

> A rapid increase in the price of food crops will intensify competition over land and other natural resources, including forest reserves. This will pit peasant farmers and indigenous communities of forest dwellers against massive agribusiness corporations and large investors who are already buying up large swathes of land or forcing peasants off their land. There could also be impacts upon employment and conditions of work.
>
> *(2008, 21)*

In addition to the observations reported above, water scarcity must be added. Biofuel production requires substantial amounts of water, which involves diverting water from food crop production. Thus, transformations to sustainable development using renewable energy in the natural and urban environments ethically and aesthetically involve some advantages and disadvantages concerning changes in the landscape. These changes interfere with biodiversity, urbanization, and the occupation of spaces by technological installations for the production and cultivation of renewable energy-generating resources, as well as environmental and urban aesthetics, as illustrated in Figures 12.1 and 12.2 with wind and solar energies.

This whole process of transformation is aimed at sustainable development. It involves a complexity of responsibilities for urban projects in design and architecture, especially for urban design planning's long-term goals, roadmaps, and

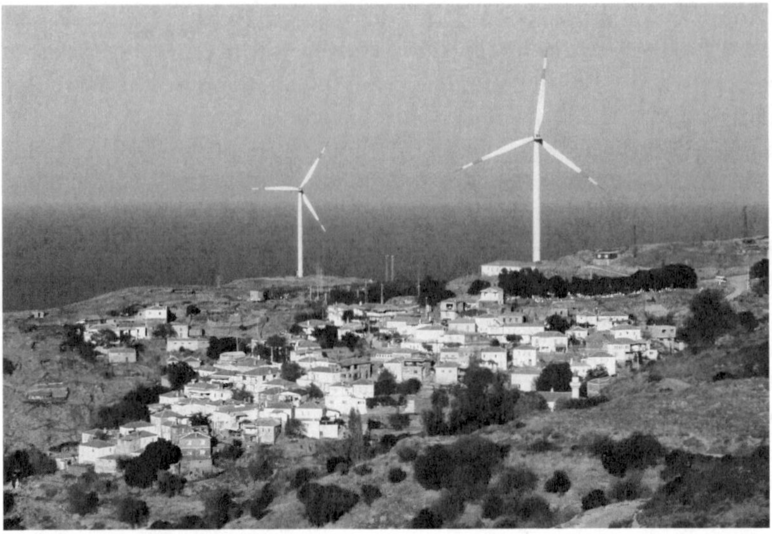

FIGURE 12.1 A small town with wind power.

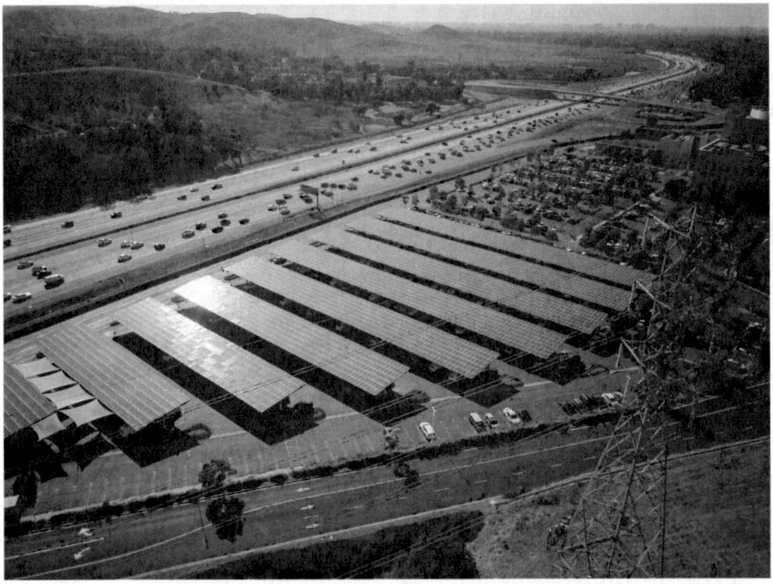

FIGURE 12.2 Landscape, city, and solar power.

management of the transition process. This planning depends on the ethical aspects of public policies and the population's needs. Nevertheless, everyone must be committed to the SDGs and the targets against global warming. In addition, the aesthetic aspects are more related to the characteristics of the environment so that architectural or urban design projects are well implemented, following the orientation of sunlight and wind. The works must follow adequate heights and positioning, considering the inclusion of green areas with trees. In addition to these aspects, according to Avi Friedman in *Fundamentals of Sustainable Urban Design*,

> a multimodal approach to network planning is a way to ensure that the system is well integrated with other modes of mobility. Public transit should also be designed in harmony with existing roads and paths to synchronize the activities on them.
>
> *(2021, 187)*

The different modes of transportation should be considered, including collective active transit and how they are organized to not only meet convenience but also offer alternative forms of transportation to the population. Inclusively, collective public transit should be prioritized to reduce individual car use to contribute to conscious consumption—energy consumption in this case. The automobile is the central element of the transformations of public space, collective life, and finally of the consequences of urbanization. With the growth of purchasing power, capital cities, and even other cities in their urbanization process did not have their ways of life modified by urbanization. The relationships with space, public infrastructure, and all of the significant and valuable objects produced to meet everyday needs are considered. As discussed earlier, the responsibility of design, a process of planning the objectivity of strategies, is also a function of urbanization. For example, the search for alternative, nonpolluting transportation through public bicycle systems can be included in the urbanization of cities with large flat areas. Friedman further explained as follows:

> Street design plays an important role in making all public transportation accessible, convenient and attractive. Streets occupy more than 40% of the city's public space and should be designed for individuals to easily and safely walk, cycle, or use public transit.
>
> *(190)*

For planning ethically and aesthetically aimed solutions, the focus is not on renovation projects, adaptation, and the transformation of urban centers to cope with global warming. Instead, it is necessary to include a minimum area distribution of green spaces, ventilation and waste heat management, trees, cool and reflective surfaces that prevent heat absorption, and the use of buildings to increase shadows in public areas, according to the manual of COP26.

In an extraordinary and complementary meeting on Sustainable Urbanization and the Implementation of the New Urban Agenda, a discussion was held regarding how the UN system can best support countries in implementing the New Urban Agenda. The said agenda comprises the strategic planning of urban spaces, establishing a new standard for sustainable urban development. For this planning to be successful, it is necessary to (1) "provide basic services for all citizens," which concerns housing, clean water and sanitation, healthy food, health and family planning, education, culture, and access to communication technologies; (2) "ensure that all citizens have access to equal opportunities and face no discrimination, as already discussed"; (3) "promote measures that support cleaner cities"—such as combating air pollution through the actions discussed above; (4) "strengthen resilience in cities to reduce the risk and the impact of disasters," which comprises general improvements in urban planning and infrastructure to reduce the risk of catastrophes; (5) "take action to address climate change by reducing their greenhouse gas emissions" according to the conditions presented above; (6) "fully respect the rights of refugees, migrants and internally displaced persons regardless of their migration status"; (7) "improve connectivity and support innovative and green initiatives"; and (8) "promote safe, accessible and green public spaces" (The New Urban Agenda 2016).

How these goals will be achieved depends on many factors discussed throughout this book's chapters. Nevertheless, it is the story of the world at its current stage, on its way to a destiny that depends on leadership, good political will, and investments in education, health, science, and technologies. The utopia of many urban planners and architects—that their concepts would change ways of life—is found in many creative designs. However, the reality consists of the concrete and plausible facts that urbanization does not create life forms but offers support for the possibilities of some models that meet the expectations of public life. The projects of urbanization, design, and architecture that meet the social reality focus on the problems generated by energy in the city. However, the issues of life and its new guidelines focus on the problems of urbanization as they relate to metropolises, in their overpopulation, behaviors, transportation, and housing. Critical aspects throughout development have already been much discussed and been part of the narratives of everyday urban life as they have been realized in urban centers.

This worldview based on the goals and guidelines of the UN represents the *Zeitgeist* of the moment with the notion that nature has its limits. Exceeding these limits puts Planet Earth at risk. Therefore, it is the leitmotif for our awareness and duty to educate those who still do not have this notion, as none of the SDGs can be achieved without everyone's commitment. In thinking about the *Zeitgeist*, as mentioned earlier, this term is related to the work of the 19th-century German philosopher Georg W. F. Hegel. It, therefore, belongs to the development of Western thought. However, to think about the *Zeitgeist* is to think about the relationship of human beings, not only in their respective societies and cultures at a specific time but also in the time–space relationship. Thus, it is essential to remember that our way of conceiving time-space may vary from culture to culture, especially between Western and non-Western countries. However, no difference exists between Western countries in this sense. Thus, specifically concerning the main characteristics that represent sociocultural transformations, the logic of Western thought, science, and technology are considered the primary motivators of the spirit of the time.

Indeed, considering the transforming and shaping powers of the *Zeitgeist*, the differences between the leading nations in their current moments of scientific and technological development are found, especially concerning industry, given the new guidelines for what is understood as sustainable development. For example, Brazil's research and solutions for renewable energy sources have been heavily invested in, especially in implementing wind power and biofuels. Professor Marcos Buckeridge of the Institute of Biosciences and coordinator of the Global Cities program of the Institute for Advanced Studies of the University of São Paulo clarified that Brazil achieved significant results in the use of renewable energy sources (Nunes 2018), thus contributing to the reduction of global warming.

Talking about the current *Zeitgeist* implies not only relating the most significant events and achievements of our contemporaneity to the term, meaning "spirit of the time." It is also necessary to clarify not only the term itself but also, and above all, the logic of how to think about the spirit of our time. Although there are many references in the history of Western thought, this term relates to Hegelian philosophy for a historical logic, as already clarified. Each moment in social history, through lived experiences, considering ideas, understandings, thoughts, wills, purposes, and cultural, ethical, and aesthetic values, achievements in their formal and universal diversities represent the spirit of the time. *Zeitgeist* concerns contemporary perception; however, it does not involve a social abstraction of the present time. On the contrary, it involves present-time participation under a relationship and understanding of the context, valuing the past. The past transmits experiences to the current one positively or negatively; thus, past experiences are recognized in their dialectical process as development in search of knowledge about the social reality and, mainly, in its evolution.

In this sense, one can consider that the main characteristic for defining the current *Zeitgeist* is perception as a synthesis of this dialectical process and transformation of our societies in a universal manner. In my view, today and since our earliest civilizations, the main characteristics of this transformative process are acquired techniques (skills), science, technology, and the material conditions that make them possible and that are strictly related to economics and politics. More precisely, today, these characteristics represent the digital world, involving all of the dynamics of communication, robotics, and nanotechnology, changing habits, and conditioning individuals in a new phase of their existence. This process undoubtedly represents a new stage of successive events that have shaped the current *Zeitgeist*, the development of capitalism, the Industrial Revolutions, science, and technology. The importance of the meaning of today's culture and society has art as a powerful reference through its themes, forms, and functions because it is in the arts that we find the representative traces of our social history, beliefs, and cultures over time. However, the present experience of all of the transformations dissolves in all aspects of daily life. These experiences transform the models of representation. This means that capturing the *Zeitgeist* is always a perception of the past. When a film, song, or artwork in any format is realized, it will represent the past, especially given the dynamics of technological evolution in communication and information in a globalized world. As such, the current *Zeitgeist* would still be seen soon. The novelty has already taken place with all successive sociocultural transformations as phenomena.

Notes

1 See UNESCO. "Anthropocene: the vital challenges of a scientific debate," last modified July 2, 2022. https://en.unesco.org/courier/2018-2/anthropocene-vital-challenges-scientific-debate.
2 Ibid.
3 Ibid.
4 See United Nations Conference on Environment and Development, Rio de Janeiro, Brazil, 3–14 June 1992. https://www.un.org/en/conferences/environment/rio1992.
5 David Nathaniel Berger and others, eds.:

> Indigenous peoples are custodians of 80 per cent of the world's remaining biodiversity, but are facing severe food insecurity, extreme poverty and other human rights deprivations. Despite the protections set forth in the United Nations Declaration on the Rights of Indigenous Peoples, mining projects, hydroelectric developments, the creation of national parks and the designation of protected areas have compromised the rights of indigenous peoples in Argentina, Bangladesh, Botswana, Canada, Chile, Ecuador, Ethiopia, Namibia, the Russian Federation, South Africa, Uganda and elsewhere.
>
> *The Indigenous World 2019* (Copenhagen, International Work Group for Indigenous Affairs, 2019): 11–14

6 Jean Ziegler: "Promotion and Protection of all Human Rights, Civil, Political, Economic, Social and Cultural Rights, Including the Right to Development." *Report of the Special Rapporteur on the right to food*, United Nations, Human Rights Council, Seventh session, Agenda item 3, 2008, 21. http://www.righttofood.org/wp-content/uploads/2012/09/AHRC75.pdf.
7 Raymond Colitt: "The growing criticism has placed Brazil at the center of the global biofuels debate. The country has enjoyed an agricultural export boom, which has seen it become the world's largest exporter of ethanol—derived from sugar cane." Critics say the increased production of crops for ethanol and biodiesel, which is derived from oil seeds, competes with for land with food crops. [...] Lula, a former union leader, rebuffed accusations by Jean Ziegler, U.N. special rapporteur for the right to food. Ziegler this week called biofuels a "crime against humanity," though he referred mainly to U.S. ethanol derived from corn. "The real crime against humanity is to discredit biofuels a priori and condemn food-starved and energy-starved countries to dependence and insecurity." "Brazil Lula defends biofuels from growing criticism," *Reuters*, April 17, 2008. https://www.reuters.com/article/us-brazil-un-biofuels-idUSN1647005420080417.
8 Jean Ziegler: "Promotion and Protection of all Human Rights, Civil, Political, Economic, Social and Cultural Rights, Including the Right to Development." *Report of the Special Rapporteur on the right to food*, United Nations, Human Rights Council, Seventh session, Agenda item 3, 2008, 21. http://www.righttofood.org/wp-content/uploads/2012/09/AHRC75.pdf.

Bibliography

Agenda 21. Accessed June 22, 2022. https://sustainabledevelopment.un.org/content/documents/Agenda21.pdf

Bégouën, Henri. "Les bases magiques de l'art préhistoriques" [The Magical Bases of Prehistoric Art]. *Scientia* 23 (1939): 202–216.

Benjamin, Walter. "Traumkitsch" [Dreamkitsch]. Originally published as "Glosse zum Sürrealismus" [Gloss on Sürrealism]. In *Die Neue Rundschau*, 1927. Reprint, *Gesammelte*

Schriften, edited by Rolf Tiedemann and Hermann Schepphäuser, *Aufsätze, Essays, Vorträge*, 620–622. Frankfurt am Main: Suhrkamp, 1977.

Berleant, Arnold. "Aesthetic Participation and the Urban Environment." *Urban Resources* 1 (1984): 37–42.

Burke, Edmund. *A Philosophical Enquiry into the Sublime and Beautiful.* 1757. Reprint, London and New York: Routledge, 2008.

Carlson, Allen. *Nature and Landscape: An Introduction to Environmental Aesthetics.* New York: Columbia University Press, 2008.

Carlson, Allen. "Contemporary Environmental Aesthetics and the Requirements of Environmentalism." *Environmental Values* 19 (2010): 289–314.

Carlson, Allen. "Environmental Aesthetics." *The Stanford Encyclopedia of Philosophy* (Winter 2020), Edward N. Zalta. https://plato.stanford.edu/archives/win2020/entries/environmental-aesthetics/

Colitt, Raymond. "Brazil Lula Defends Biofuels from Growing Criticism." *Reuters*, April 17, 2008.

Dewey, John. *Experience and Nature.* New York: Dover, 1925.

Elkington, John. *Cannibals with Forks: The Triple Bottom Line of 21st Century Business.* Oxford: Capstone Publishing Limited, Oxford Centre for Innovation, 1997.

Friedman, Avi. *Fundamentals of Sustainable Urban Design.* Cham: Springer, 2021.
https://doi.org/10.1007/978-3-030-60865-1_20

Hegel, Georg Wilhelm Friedrich. *The Phenomenology of Spirit.* Trans. by Terry Pinkard (Cambridge Hegel Translations). Cambridge: University Press, 2018. Originally published as *Die Phänomenologie des Geistes* (Bamberg and Würzburg: Joseph Anton Goebhardt, 1807).

Hegel, Georg Wilhelm Friedrich. *Lectures on the Philosophy of Art.* The Hotho Transcript of the 1823 Berlin Lectures. Trans. by R.F. Brown, Oxford: Clarendon Press, 2014. Originally published as *Vorlesung über die Philosophie der Kunst.* Berlin 1823. Reprinted by Heinrich Gustav Hotho (Hamburg: Felix Meiner, 1998).

Kant, Immanuel. *Kritik der Urteilskraft* [Critique of the Power of Judgment]. Berlin and Libau: Lagarde und Friedrich 1790. Reprint, Stuttgart: Reclam, 1963.

Löwy, Michael. "Walter Benjamin's Romantic Utopia." *Fênix, Revista de História e Estudos Culturais* 16 (2020): 1–16. https://doi.org/10.35355/0000021.

Morton, Timothy. *Ecology without Nature: Rethinking Environmental Aesthetics.* Cambridge, MA and London: Harvard University Press, 2007.

Nunes, Lucas. "Brasil tem alta taxa de energia renovável na matriz energética" [Brazil Has a High Rate of Sustainable Power in the Energy Matrix]. *Jornal da USP*, June 4, 2018. https://jornal.usp.br/?p=170944

Recki, Birgit. "Natur und Technik. Eine Komplikation" [Nature and Technology. A Complication]. In *De Natura* VIII, edited by Frank Fehrenbach. Berlin: Matthes und Seitz, 2021.

Rosny, J.-H. *The Quest for Fire.* New York: Pantheon Books, 1911.

Schmitz, Alfried. "Die Romantik—Epoche der Sehnsucht" [Romanticism—Epoch of Nostalgia]. *Planet Wissen*, January 21, 2008. http://www.planetwissen.de/kultur_medien/literatur/literatur_der_romantik/wissensfrage_blaue_blume.jsp

The New Urban Agenda. https://habitat3.org/the-new-urban-agenda/

United Nations. "It's 'Now or Never' to Limit Global Warming to 1.5 Degrees." *UN Climate Report, Urban Issue.* Accessed June 22, 2022. https://news.un.org/en/story/2022/04/1115452

UN Climate Change Conference. "The 21st Conference of the Parties (COP21)." Accessed June 22, 2022. https://www.un.org/sustainabledevelopment/cop21/

Welsch, Wolfgang. "Animal Aesthetics." *Contemporary Aesthetics* 2 (2004): 1–31. https://digitalcommons.risd.edu/liberalarts_contempaesthetics/vol2/iss1/15/

Ziegler, Jean. *Report of the Special Rapporteur on the right to food.* United Nations, Human Rights Council, Seventh session, Agenda item 3, (2008): 21. http://www.righttofood.org/wp-content/uploads/2012/09/AHRC75.pdf.

INDEX

For Product Safety Concerns and Information please contact our EU representative GPSR@taylorandfrancis.com Taylor & Francis Verlag GmbH, Kaufingerstraße 24, 80331 München, Germany